Natural Environment Research Council

Dr. P.S. Maitland

Institute of Terrestrial Ecology

A Coded Checklist
of Animals
occurring in Fresh Water
in the
British Isles.

First published 1977
Institute of Terrestrial Ecology
c/o Nature Conservancy Council
12 Hope Terrace
EDINBURGH
EH9 2AS
031 447 (Edinburgh) 4784

The Institute of Terrestrial Ecology (ITE) was established in 1973, from the former Nature Conservancy's research stations and staff, joined later by the Institute of Tree Biology and the Culture Centre of Algae and Protozoa. ITE contributes to and draws upon the collective knowledge of the fourteen sister institutes which make up the Natural Environment Research Council, spanning all the envrionmental sciences.

The Institute studies the factors determining the structure, composition and processes of land and freshwater systems, and of individual plant and animal species. It is developing a sounder scientific basis for predicting and modelling environmental trends arising from natural or man-made change. The results of this research are available to those responsible for the protection, management and wise use of our natural resources.

Nearly half of ITE's work is research commissioned by customers, such as the Nature Conservancy Council who require information for wildlife conservation, the Forestry Commission and the Department of the Environment. The remainder is fundamental research supported by NERC.

ITE's expertise is widely used by international organisations in overseas projects and programmes of research.

Introduction

The purpose of this publication is to provide a comprehensive list of all free-living animals, from sponges to mammals but excluding Protozoa, which occur in, or in association with, freshwater systems in the British Isles. It is organised in such a way that each species can be represented by a unique 8-digit code, thus establishing a standard and relatively easy basis for handling species lists, etc. numerically. This check list, containing over 3,800 species has been produced in parallel to a coded list of all lakes in Great Britain shown on 1: 250,000 maps (ca 5,600 lakes) recently completed by Smith (1976). The two studies, together with other proposed projects (eg. coded lists of algae, macrophytes and rivers) provide the basis of a powerful tool for handling and analysing data on freshwater ecosystems in this country.

As far as the actual species listed are concerned it has been necessary to make a number of judgement decisions regarding those to be included or excluded. In general, where a recent taxonomic key or check-list to freshwater species in the British Isles is available for any group then this has been followed more or less exactly. Where there is doubt about the validity of records of a species it has normally been included in the list to avoid possible insertion at a later date. Thus the presence in the list of any particular species should not be taken as an assurance that it occurs in the British Isles. As far as habitat is concerned similar decisions have been made where there is doubt as to whether any species is aquatic or terrestrial, freshwater or marine. Thus parallel check lists of terrestrial or marine animals will be likely to contain some of the animals listed here.

It is intended that this list will be useful to those freshwater ecologists throughout the country who are concerned with sampling, identifying and analysing mixed collections of animals. Thus it is closely linked with the taxonomic keys in greatest use. Some problems have arisen, however, related to keys which do not have an accompanying check list or where new species have been discovered since publication. In these cases compromises have been reached based on the best literature and advice available. The main published sources of information used for each group are indicated in Table 1: this also includes data on the number of species and the location of groups within the list.

The format of the check list itself is quite simple and the numbers meaningful taxonomically. The first pair of digits represents the major groupings - usually phyla, classes or orders. The 2nd, 3rd and 4th pairs of digits represent families (occasionally subfamilies), genera and species (with authorities) respectively. The numbering of the check list follows a logical sequence and space is available for the addition of new species.

The present list has a number of advantages in addition to the obvious one of being able to represent any species by only 8 digits. The great majority of lists of animals from freshwater habitats in this country are incomplete as far as identification is concerned and many groups (e.g. Chironomidae) are rarely identified beyond family. The present system is completely flexible in such situations for while a full identification of *Chlorohydra viridissima* (Pallas) is coded 02010201 if identification is taken no further than the family (Hydridae) the coding is 02010000. Thus any form of taxonomic list of animals can be coded relatively easily.

The increasing numbers of studies of aquatic animals being carried out by freshwater ecologists in many parts of the country, especially by those working in the fields of water supply, pollution prevention, fisheries and conservation, mean that acceptance of a common check list, particularly a coded one, is becoming more and more desirable. It is suggested that the present list is extensive and versatile enough to meet the needs of almost all freshwater ecologists in the British Isles.

In addition to relying heavily on the published literature, the author has been helped greatly in the production of this list by general criticism or comments on various groups from the following people: A. Brindle, J.C. Chubb, R.A. Crowson, J.T. Dealtry, J.M. Elliot, G. Fryer, D.C. Geddes, P.S. Gooch, R. Gibson, J. Green, J.D. Hamilton, D.J. Hooper, J.P.K. Horkan, J.N.R. Jeffers, D. Jenkins, R. Jones, C.R. Kennedy, E.D. Le Cren, R.J. Lincoln, T.T. Macan, R.M. Pontin, S. Prudhoe, O.W. Richards, G.S. Scott, M.W. Service, K.G.V. Smith, W.J.P. Smyly, S.M. Stone, D.W. Sutcliffe, and J.O. Young. The coding was carried out by Dr C.F. Mason. Mrs. J. Garcia, Miss M. MacDonald and Mrs S. Adair helped greatly with the initial preparation of the list and Mrs M.S. Wilson had the unenviable job of typing it. I am grateful to everyone for their help.

Table 1.   The major groups of animals included in the coded check list, with total numbers of species and major references.

CODE	MAJOR GROUPS		TOTAL SPECIES	PAGE	REFERENCES
01	PORIFERA		8	7	86,101
02	COELENTERATA		8	7	42,96
03	PLATYHELMINTHES	TURBELLARIA: MICROTURBELLARIA	46	7	118,114
		TURBELLARIA: TRICLADIDA	11	8	91
04		TREMATODA	*	8	23
05		CESTODA	*	8	112
06	NEMERTINI		2	8	35,36
07	NEMATOMORPHA		4	8	4,93
08	ROTIFERA		511	8	6,51,88,94,109
09	ACANTHOCEPHALA		*	17	52
10	NEMATODA		97	17	1,41
11	GASTROTRICHA		22	19	90,110
12	POLYZOA		9	19	65,111
13	MOLLUSCA	GASTROPODA	52	19	27,73,75
14		BIVALVIA	27	20	27,28
15	ANNELIDA	POLYCHAETA	1	21	25
16		OLIGOCHAETA	118	21	7,8,34,81,102
17		HIRUDINEA	14	23	77
18	ARTHROPODA: OTHERS	TARDIGRADA	35	23	67,80,89
19		HYDRACARINA	322	24	39,100
20		ORIBATEI	4	29	3,87
21		ARANEAE	1	29	68
22	ARTHROPODA: CRUSTACEA	ANOSTRACA	2	29	70
23		NOTOSTRACA	1	29	70

CODE	MAJOR GROUPS		TOTAL SPECIES	PAGE	REFERENCES
24		CLADOCERA	88	29	97
25		OSTRACODA	88	31	11,69
26		COPEPODA	111	33	43,46,66
27		BRANCHIURA	2	34	44,58
28		MALACOSTRACA	33	35	40
29	ARTHROPODA: INSECTA	COLLEMBOLA	17	35	37,38,64
30		EPHEMEROPTERA	47	35	61,71
31		PLECOPTERA	34	36	53
32		ODONATA	45	37	19,30,64
33		HEMIPTERA	62	38	72,64
34		HYMENOPTERA	27	39	47,57,92
35		COLEOPTERA	300	39	2,50,55
36		MEGALOPTERA	2	45	59
37		NEUROPTERA	4	45	59
38		TRICHOPTERA	193	45	49,60,74
39		LEPIDOPTERA	11	48	45,63,79
40		DIPTERA	1138	48	54,64,82
41	VERTEBRATA	AGNATHA	3	68	76
		PISCES	51	68	76
42		AMPHIBIA	8	69	99
43		REPTILIA	2	69	99
44		AVES	205	69	9,10,29
45		MAMMALIA	8	73	18

* *All members of these groups are internal parasites and not covered in this list.*

Table 2. The families of Diptera included in the coded check list, with total number of species and major references.

INSECTA				TOTAL SPECIES	PAGE	REFERENCES
40	DIPTERA	01	Tipulidae	117	48	13,64
		02	Psychodidae	61	50	32,64
		03	Ptychopteridae	7	51	31,64
		04	Dixidae	14	52	24
		05	Chaoboridae	6	52	33,64,95
		06	Culicidae	33	52	64,78
		07	Thaumaleidae	3	52	26,64
		08	Ceratopogonidae	153	52	12,64
		09	Chironomidae	450	55	14,64
		15	Simuliidae	35	63	22
		16	Stratiomyidae	27	63	64,107
		17	Empididae	48	64	17,103
		18	Dolichopodidae	59	64	84,104
		19	Rhagionidae	1	65	56,64
		20	Tabanidae	12	65	56,64,83
		21	Syrphidae	26	66	15,105
		22	Ephydridae	32	66	5,20
		23	Sciomyzidae	43	67	85,106
		24	Scatophagidae	4	67	16,64,108
		25	Muscidae	7	68	21,48

Coded Check List

01 PORIFERA   DEMOSPONGIAE

Spongillidae	*Spongilla*	*lacustris*(L)	01 01 01 01
	Eunapius	*fragilis* (Leidy)	01 01 02 01
	Ephydatia	*fluviatilis* (L.)	01 01 03 01
		mulleri (Leiberkubn)	01 01 03 02
	Heteromeyenia	*baileyi* (Bowerbank)	01 01 04 01
		stepanowii (Dybowsky)	01 01 04 02
	Anheteromeyenia	*ryderi* (Potts)	01 01 05 01
	Trochospongilla	*horrida* Weltner	01 01 06 01

02 COELENTERATA   HYDROZOA

Hydridae	*Hydra*	*attenuata* Pallas	02 01 01 01
		circumcincta Schulze	02 01 01 02
		oligactis (Pallas)	02 01 01 03
		graysoni Maxwell	02 01 01 04
	Chlorohydra	*viridissima* (Pallas)	02 01 02 01
	Protohydra	*leuckarti* Greef	02 01 03 01
Clavidae	*Cordylophora*	*lacustris* Allman	02 02 01 01
Olindiidae	*Craspedacusta*	*sowerbii* Lankester	02 03 01 01

03 PLATYHELMINTHES   TURBELLARIA: MICROTURBELLARIA

Catenulidae	*Catenula*	*lemnae* Duges	03 01 01 01
Stenostomidae	*Rhynchoscolex*	*simplez* Leidy	03 02 01 01
	Stenostomum	*leucops* (Duges)	03 02 02 01
		unicolor Schmidt	03 02 02 02
		bryophilum Luther	03 02 02 03
		grabbskogense Luther	03 02 02 04
Microstomidae	*Microstomum*	*lineare* (Muller)	03 03 01 01
Macrostomidae	*Macrostomum*	*rostratum* (Papi)	03 04 01 01
		distinguendum (Papi)	03 04 01 02
		johni Young	03 04 01 03
Prorhynchidae	*Prorhynchus*	*stagnalis* Schultze	03 05 01 01
	Geocentrophora	*sphyrocephala* Man	03 05 02 01
		baltica (Kennel)	03 05 02 02
Plagiostomidae	*Plagiostomum*	*lemani* (Plessis)	03 06 01 01
Bothrioplanidae	*Bothrioplana*	*semperi* Braun	03 07 01 01
Otomesostomidae	*Otomesostoma*	*auditivum* (Plessis)	03 08 01 01
Dalyellidae	*Microdalyellia*	*fairchildi* (Graff)	03 09 01 01
		schmidtii (Graff)	03 09 01 02
		kupelwieseri (Meisner)	03 09 01 03
		brevimana (Beklemischev)	03 09 01 04
	Gieysztoria	*diadema* (Hofsten)	03 09 02 01
		infundibuliformis Fuhrmann	03 09 02 02
		rubra (Fuhrmann)	03 09 02 03
	Castrella	*truncata* (Abilgaard)	03 09 03 01
	Dalyellia	*viridis* (Shaw)	03 09 04 01
Typhloplanidae	*Typhloplana*	*viridata* (Abildgaard)	03 10 01 01
	Castrada	*luteola* Hofsten	03 10 02 01
		armata (Fuhrmann)	03 10 02 02
		lanceola Braun	03 10 02 03
		neocomensis Volz	03 10 02 04
		viridis Volz	03 10 02 05

	Tetracelis	*marmorosa* (Muller)	03 10 03 01
	Strongylostoma	*radiatum* (Muller)	03 10 04 01
		elongatum Hofsten	03 10 04 02
	Olisthanella	*obtusa* (Schultze)	03 10 05 01
	Rhynchomesostoma	*rostratum* (Muller)	03 10 06 01
	Mesostoma	*lingua* (Abildgaard)	03 10 07 01
		productum (Schmidt)	03 10 07 02
		tetragonum (Muller)	03 10 07 03
		ehrenbergii (Focke)	03 10 07 04
	Bothromesostoma	*personatum* (Schmidt)	03 10 08 01
	Phaenocora	*unipunctata* (Orsted)	03 10 09 01
		typhlops (Vejdovsky)	03 10 09 02
	Opistomum	*pallidum* Schmidt	03 10 10 01
Polycystididae	*Gyratrix*	*hermaphroditus* Ehrenberg	03 11 01 01
	Opisthocystis	*goettei* (Bresslau)	03 11 02 01

TURBELLARIA: TRICLADIDA

Planariidae	*Planaria*	*torva* (Muller)	03 12 01 01
	Polycelis	*nigra* (muller)	03 12 02 01
		tenuis (Ijima)	03 12 02 02
		felina (Dalyell)	03 12 02 03
	Dugesia	*lugubris* (Schmidt)	03 12 03 01
		tigrina (Girard)	03 12 03 02
		polychroa (Schmidt)	03 12 03 03
	Phagocata	*vitta* (Duges)	03 12 04 01
	Crenobia	*alpina* (Dana)	03 12 05 01
Dendrocoelidae	*Dendrocoelum*	*lacteum* (Muller)	03 13 01 01
	Bdellocephala	*punctata* (Pallas)	03 13 02 01

04 TREMATODA			04 00 00 00
05 CESTODA			05 00 00 00
06 NEMERTINI	ENOPLA		
Tetrastemmidae	*Prostoma*	*graecense* (Bohmig)	06 01 01 01
		jenningsi Gibson & Young	06 01 01 02
07 NEMATOMORPHA	GORDIOIDEA		
Chordodidae	*Gordius*	*villoti* Rosa	07 01 01 01
	Parachordodes	*violaceus* (Baird)	07 01 02 01
		wolterstorffi (Camerano)	07 01 02 02
		pustulosus (Baird)	07 01 02 03
08 ROTIFERA	BDELLOIDEA		
Habrotrochidae	*Otostephanus*	*torquatus* (Bryce)	08 01 01 01
	Scephanotrocha	*corniculata* Bryce	08 01 02 01
		rubra Bryce	08 01 02 02
	Habrotrocha	*angusticollis* (Murray)	08 01 03 01
		annulata (Murray)	08 01 03 02
		appendiculata Murray	08 01 03 03
		aspera (Bryce)	08 01 03 04
		bidens (Gosse)	08 01 03 05
		collaris (Ehrenberg)	08 01 03 06
		constricta (Dujardin)	08 01 03 07
		crenata Murray	08 01 03 08
		elegans (Milne)	08 01 03 09
		eremita (Bryce)	08 01 03 10
		flava Bryce	08 01 03 11
		flaviformis de Koning	08 01 03 12
		fusca (Bryce)	08 01 03 13
		insignis Bryce	08 01 03 14

		lata (Bryce)	08 01 03 15
		leitgebi (Zelinka)	08 01 03 16
		ligula Bryce	08 01 03 17
		longiceps (Murray)	08 01 03 18
		longula Bryce	08 01 03 19
		microcephala (Murray)	08 01 03 20
		minuta (Murray)	08 01 03 21
		munda Bryce	08 01 03 22
		pavida Bryce	08 01 03 23
		pulchra (Murray)	08 01 03 24
		pusilla (Bryce)	08 01 03 25
		reclusa (Milne)	08 01 03 26
		roeperi (Milne)	08 01 03 27
		spicula Bryce	08 01 03 28
		sylvestris Bryce	08 01 03 29
		thermalis Pax & Wulfert	08 01 03 30
		tridens (Milne)	08 01 03 31
		tripus (Murray)	08 01 03 32
		visa Donner	08 01 03 33
Adinetidae	*Adineta*	*barbata* Janson	08 02 01 01
		gracilis Janson	08 02 01 02
		oculata (Milne)	08 02 01 03
		tuberculosa Janson	08 02 01 04
		vaga (Davis)	08 02 01 05
	Bradyscela	*clauda* (Bryce)	08 02 02 01
Philodinidae	*Mniobia*	*armata* (Murray)	08 03 01 01
		circinata (Murray)	08 03 01 02
		incrassata (Murray)	08 03 01 03
		magna (Plate)	08 03 01 04
		russeola (Zelinka)	08 03 01 05
		scarlatina (Ehrenberg)	08 03 01 06
		symbiotica (Zelinka)	08 03 01 07
		tetraodon (Ehrenberg)	08 03 01 08
	Ceratotrocha	*cornigera* (Bryce)	08 03 02 01
	Rotaria	*citrina* (Ehrenberg)	08 03 03 01
		curtipes (Murray)	08 03 03 02
		elongata (Weber)	08 03 03 03
		haptica (Gosse)	08 03 03 04
		macroceros (Gosse)	08 03 03 05
		macrura (Ehrenberg)	08 03 03 06
		magna-calcarata Parsons	08 03 03 07
		murrayi Bartos	08 03 03 08
		neptunia (Ehrenberg)	08 03 03 09
		neptunoida Harring	08 03 03 10
		quadrioculata (Murray)	08 03 03 11
		rotatoria (Pallas)	08 03 03 12
		socialis (Kellicott)	08 03 03 13
		sordida (Western)	08 03 03 14
		spicata (Murray)	08 03 03 15
		tardigrada (Ehrenberg)	08 03 03 16
		trisecata (Weber)	08 03 03 17
	Macrotrachela	*aculeata* Milne	08 03 04 01
		angusta (Bryce)	08 03 04 02
		asperula (Murray)	08 03 04 03
		bilfingeri (Bryce)	08 03 04 04
		bullata (Murray)	08 03 04 05
		concinna (Bryce)	08 03 04 06
		crucicornis (Murray)	08 03 04 07
		decora (Bryce)	08 03 04 08
		ehrenbergi (Janson)	08 03 04 09
		fungicola Garn.	08 03 04 10
		habita (Bryce)	08 03 04 11
		multispinosa Thompson	08 03 04 12

		muricata (Murray)	08 03 04 13
		musculosa Milne	08 03 04 14
		nana (Bryce)	08 03 04 15
		natans (Murray)	08 03 04 16
		ornata Donner	08 03 04 17
		papillosa Thompson	08 03 04 18
		plicata (Bryce)	08 03 04 19
		quadricornifera Milne	08 03 04 20
		vesicularis (Murray)	08 03 04 21
	Embata	*commensalis* (Western)	08 03 05 01
		hamata (Murray)	08 03 05 02
		laticeps (Murray)	08 03 05 03
		laticornis (Murray)	08 03 05 04
		parasitica (Giglioli)	08 03 05 05
	Philodina	*acutinornis* Murray	08 03 06 01
		brevipes Murray	08 03 06 02
		citrina Ehrenberg	08 03 06 03
		convergens Murray	08 03 06 04
		erythrophthalma Ehrenberg	08 03 06 05
		flaviceps Bryce	08 03 06 06
		megalotrocha Ehrenberg	08 03 06 07
		nemoralis Bryce	08 03 06 08
		plena (Bryce)	08 03 06 09
		roseola Ehrenberg	08 03 06 10
		rugosa Bryce	08 03 06 11
		vorax (Janson)	08 03 06 12
	Dissotrocha	*aculeata* (Ehrenberg)	08 03 07 01
		macrostyla (Ehrenberg)	08 03 07 02
		spinosa (Bryce)	08 03 07 03
	Pleuretra	*alpium* (Ehrenberg)	08 03 08 01
		brycei (Weber)	08 03 08 02
		humerosa (Murray)	08 03 08 03
Philodinavidae	*Philodinavus*	*paradoxus* (Murray)	08 04 01 01
	Henoceros	*falcatus* Milne	08 04 02 01
	MONOGONONTA		
Brachionidae	*Proalides*	*tentaculatus* Beauchamp	08 05 01 01
	Micrococides	*chlaena* (Gosse)	08 05 02 01
		robustus (Glascott)	08 05 02 02
	Cyrtonia	*tuba* (Ehrenberg)	08 05 03 01
	Rhinoglena	*frontalis* Ehrenberg	08 05 04 01
	Epiphanes	*brachionus* (Ehrenberg)	08 05 05 01
		clavulata (Ehrenberg)	08 05 05 02
		senta (Muller)	08 05 05 03
	Macrochaetus	*subquadratus* Perty	08 05 06 01
		collinsi (Gosse)	08 05 06 02
	Trichotria	*pocillum* (Muller)	08 05 07 01
		tetractis (Ehrenberg)	08 05 07 02
		truncata (White)	08 05 07 03
	Platyias	*patulus* (Muller)	08 05 08 01
		quadricornis (Ehrenberg)	08 05 08 02
	Brachonius	*angularis* Gosse	08 05 09 01
		calyciflorus Pallas	08 05 09 02
		leydigii Cohn	08 05 09 03
		plicatilis Muller	08 05 09 04
		quadridentatus Hermann	08 05 09 05
		rubens Ehrenberg	08 05 09 06
		urceolaris Muller	08 05 09 07
	Eudactylota	*eudactylota* (Gosse)	08 05 10 01

Wolga	*spinifera* (Wester)	08 05 11 01
Lophocharis	*oxysternum* (Gosse)	08 05 12 01
	salpina (Ehrenberg)	08 05 12 02
Mytilina	*compressa* (Gosse)	08 05 13 01
	crassipes (Lucks)	08 05 13 02
	mucronata (Muller)	08 05 13 03
	mutica (Perty)	08 05 13 04
	trigona (Gosse)	08 05 13 05
	unguipes (Lucks)	08 05 13 06
	ventralis (Ehrenberg)	08 05 13 07
Diplois	*daviesiae* Gosse	08 05 14 01
Tripleuchlanis	*plicata* (Levander)	08 05 15 01
Euchlanis	*deflexa* Gosse	08 05 16 01
	dilatata Ehrenberg	08 05 16 02
	incisa Carlin	08 05 16 03
	lyra Hudson	08 05 16 04
	meneta Myers	08 05 16 05
	parva Rousselet	08 05 16 06
	proxima Myers	08 05 16 07
	pyriformis Gosse	08 05 16 08
	triquetra Ehrenberg	08 05 16 09
Dipleuchlanis	*propatula* (Gosse)	08 05 17 01
Anuraeopsis	*fissa* Lauterborn	08 05 18 01
Keratella	*cochlearis* (Gosse)	08 05 19 01
	cruciformis (Thompson)	08 05 19 02
	hiemalis Carlin	08 05 19 03
	quadrata (Muller)	08 05 19 04
	serrulata (Ehrenberg)	08 05 19 05
	testudo (Ehrenberg)	08 05 19 06
	ticinensis (Callerio)	08 05 19 07
	valga (Ehrenberg)	08 05 19 08
Argonotholca	*foliacea* (Ehrenberg)	08 05 20 01
Kellicottia	*longispina* (Kellicott)	08 05 21 01
Notholca	*acuminata* (Ehrenberg)	08 05 22 01
	cinetura Skorikov	08 05 22 02
	labis Gosse	08 05 22 03
	squamula (Muller)	08 05 22 04
	striata (Muller)	08 05 22 05
Squatinella	*bifurca* (Bolton)	08 05 23 01
	leydigii (Zacharias)	08 05 23 02
	longispinata (Tatem)	08 05 23 03
	microdactyla (Murray)	08 05 23 04
	mutica (Ehrenberg)	08 05 23 05
	rostrum (Schmarda)	08 05 23 06
	tridentata (Fresenius)	08 05 23 07
Lepadella	*acuminata* (Ehrenberg)	08 05 24 01
	cristata (Rousselet)	08 05 24 02
	cryphaea Harring	08 05 24 03
	ehrenbergi (Perty)	08 05 24 04
	minuta (Montet)	08 05 24 05
	oblonga (Ehrenberg)	08 05 24 06
	ovalis (Muller)	08 05 24 07
	patella (Muller)	08 05 24 08
	pterygoides (Dunlop)	08 05 24 09
	rhomboides (Gosse)	08 05 24 10
	rhonboidula (Bryce)	08 05 24 11
	triptera Ehrenberg	08 05 24 12
Colurella	*adriatica* Ehrenberg	08 05 25 01
	colurus (Ehrenberg)	08 05 25 02

		dicentra (Gosse)	08 05 25 03
		gastracantha Hauer	08 05 25 04
		linderburgi Steinecke	08 05 25 05
		obtusa (Gosse)	08 05 25 06
		sinistra Carlin	08 05 25 07
		tessellata (Glascott)	08 05 25 08
		uncinata (Muller)	08 05 25 09
	Paracolurella	pertyi (Hood)	08 05 26 01
Lecanidae	*Lecane*	acuminata (Ehrenberg)	08 06 01 01
		acus Harring	08 06 01 02
		agilis (Bryce)	08 06 01 03
		aquila Harring & Myers	08 06 01 04
		arcuata (Bryce)	08 06 01 05
		bifurca (Bryce)	08 06 01 06
		brachydactyla (Stenroos)	08 06 01 07
		bulla (Gosse)	08 06 01 08
		clara (Bryce)	08 06 01 09
		closterocerca (Schmarda)	08 06 01 10
		cornuta (Muller)	08 06 01 11
		depressa (Bryce)	08 06 01 12
		diomis (Gosse)	08 06 01 13
		flexilis (Gosse)	08 06 01 14
		furcata (Murray)	08 06 01 15
		galeata (Bryce)	08 06 01 16
		hamata (Stokes)	08 06 01 17
		hornemanni (Ehrenberg)	08 06 01 18
		inermis (Bryce)	08 06 01 19
		intrasinuata (Olofssen)	08 06 01 20
		latifrons (Gosse)	08 06 01 21
		ligona Dunlop	08 06 01 22
		ludwigi Eckstein	08 06 01 23
		luna (Muller)	08 06 01 24
		lunaris (Ehrenberg)	08 06 01 25
		ploenensis (Voigt)	08 06 01 26
		pumila (Rousselet)	08 06 01 27
		pyrrha Harring & Myers	08 06 01 28
		quadridentata (Ehrenberg)	08 06 01 29
		rusticula (Gosse)	08 06 01 30
		saginata Harring & Myers	08 06 01 31
		signifera Harring & Myers	08 06 01 32
		stenroosi (Meissner)	08 06 01 33
		stichaea Harring	08 06 01 34
		stokesi (Pell)	08 06 01 35
		striata Gosse	08 06 01 36
		subulata Harring & Myers	08 06 01 37
		sulcata (Gosse)	08 06 01 38
		ungulata (Gosse)	08 06 01 39
	Bryceella	stylata (Milne)	08 06 02 01
		tenella (Bryce)	08 06 02 02
	Tetrasiphon	hydrocora Ehrenberg	08 06 03 01
	Proalinopsis	caudatus (Collins)	08 06 04 01
		squamipes Hauer	08 06 04 02
		staurus Harring & Myers	08 06 04 03
	Proales	coryneger Gosse	08 06 05 01
		daphnicola (Thompson)	08 06 05 02
		decipiens (Ehrenberg)	08 06 05 03
		doliaris (Rousselet)	08 06 05 04
		fallaciosa Wulfert	08 06 05 05
		gigantea (Glascott)	08 06 05 06
		globulifera (Hauer)	08 06 05 07
		latrunculus Penard	08 06 05 08
		micropus (Gosse)	08 06 05 09
		minima (Montet)	08 06 05 10

		parasita (Ehrenberg)	08	06	05	11
		prehensor Gosse	08	06	05	12
		reinhardti (Ehrenberg)	08	06	05	13
		similis (Beauchamp)	08	06	05	14
		sondida (Gosse)	08	06	05	15
		theodora (Gosse)	08	06	05	16
		wernecki (Ehrenberg)	08	06	05	17
Lindiidae	*Lindia*	*torulosa* Dujardin	08	07	01	01
		yanickii Wisniewski	08	07	01	02
Notommatidae	*Scaridium*	*longicaudun* (Muller)	08	08	01	01
	Monommata	*actices* Myers	08	08	02	01
		aequalis (Ehrenberg)	08	08	02	02
		astra Myers	08	08	02	02
		grandis Tessin	08	08	02	04
		longiseta (Muller)	08	08	02	05
	Dorystoma	*caudata* (Bilfinger)	08	08	03	01
	Itura	*aurita* (Ehrenberg)	08	08	04	01
		viridis (Stenroos)	08	08	04	02
	Enteroplea	*lacustris* (Ehrenberg)	08	08	05	01
	Metadiaschiza	*trigona* (Rousselet)	08	08	06	01
	Taphrocampa	*annulosa* Gosse	08	08	07	01
		clavigera Stokes	08	08	07	02
		selenura (Gosse)	08	08	07	03
	Eothinia	*elongata* (Ehrenberg)	08	08	09	01
		lucens (Glascott)	08	08	09	02
	Cephalodella	*auriculata* (Muller)	08	08	10	01
		biungulata Wulfert	08	08	10	02
		catellina (Muller)	08	08	10	03
		crassipes (Lord)	08	08	10	04
		derbyi Dixon-Nuttall & Freeman	08	08	10	05
		eva (Gosse)	08	08	10	06
		exigua (Gosse)	08	08	10	07
		forficata (Ehrenberg)	08	08	10	08
		forficula (Ehrenberg)	08	08	10	09
		gibba (Ehrenberg)	08	08	10	10
		globata (Gosse)	08	08	10	11
		gracilis (Ehrenberg)	08	08	10	12
		hoodi (Gosse)	08	08	10	13
		megalocephala (Glascott)	08	08	10	14
		myersi Wisniewski	08	08	10	15
		obvia Donner	08	08	10	16
		sterea (Gosse)	08	08	10	17
		strigosa Myers	08	08	10	18
		tenuior (Gosse)	08	08	10	19
		tenuiseta (Burn)	08	08	10	20
		ventripes Dixon-Nuttall	08	08	10	21
	Notommata	*allantois* Wulfert	08	08	11	01
		aurita (Muller)	08	08	11	02
		brachyota Ehrenberg	08	08	11	03
		cerberus (Gosse)	08	08	11	04
		collaris Ehrenberg	08	08	11	05
		contorta (Stokes)	08	08	11	06
		copeus Ehrenberg	08	08	11	07
		cyrtopus Gosse	08	08	11	08
		groenlandica Bergendal	08	08	11	09
		pachyura (Gosse)	08	08	11	10
		pavida Myers	08	08	11	11
		potamis Gosse	08	08	11	12
		pseudocerberus Beauchamp	08	08	11	13
		saccigera Ehrenberg	08	08	11	14
		silpha (Gosse)	08	08	11	15

		tripus Ehrenberg	08 08 11 16
	Resticula	*melandocus* (Gosse)	08 08 12 01
	Eosphora	*ehrenbergi* (Weber)	08 08 13 01
		gibba Garn	08 08 13 02
		najas Ehrenberg	08 08 13 03
	Pleurotrocha	*constricta* (Ehrenberg)	08 08 14 01
		petromyzon Ehrenberg	08 08 14 02
Trichoceridae	*Trichocerca*	*bicristata* (Gosse)	08 09 01 01
		bidens (Lucks)	08 09 01 02
		birostris (Minkiwicz)	08 09 01 03
		brachyura (Gosse)	08 09 01 04
		capucina (Wierzejski&Zacharias)	08 09 01 05
		cavia (Gosse)	08 09 01 06
		collaris (Rousselet)	08 09 01 07
		cylindrica (Imhof)	08 09 01 08
		dixon-nuttalli (Jennings)	08 09 01 09
		elongata (Gosse)	08 09 01 10
		helminthoides (Gosse)	08 09 01 11
		iernis (Gosse)	08 09 01 12
		inermis (Linder)	08 09 01 13
		intermedia (Stenroos)	08 09 01 14
		jenningsi Voigt	08 09 01 15
		longiseta (Schrank)	08 09 01 16
		lophoessa (Gosse)	08 09 01 17
		macera (Gosse)	08 09 01 18
		marina (Daday)	08 09 01 19
		porcellus (Gosse)	08 09 01 20
		pusilla (Jennings)	08 09 01 21
		rattus (Muller)	08 09 01 22
		rousseleti (Voigt)	08 09 01 23
		scipio (Gosse)	08 09 01 24
		sejunctipes (Gosse)	08 09 01 25
		similis (Wierzejski)	08 09 01 26
		stylata (Gosse)	08 09 01 27
		sulcata (Jennings)	08 09 01 28
		tenuior (Gosse)	08 09 01 29
		tigris (Muller)	08 09 01 30
		weberi Jennings	08 09 01 31
	Hertwigella	*volvocicola* (Plate)	08 09 02 01
	Elosa	*woralli* Lord	08 09 03 01
Gastropodidae	*Gastropus*	*hyptopus* (Ehrenberg)	08 10 01 01
		minor (Rousselet)	08 10 01 02
		stylifer Imhof	08 10 01 03
	Ascomorpha	*ecaudis* Perty	08 10 02 01
		saltans Bartsch	08 10 02 02
	Chromogaster	*ovalis* (Bergendal)	08 10 03 01
Dicranophoridae	*Albertia*	*bernardi* Hlava	08 11 01 01
		intrusor Gosse	08 11 01 02
		naidis Bousfield	08 11 01 03
	Aspelta	*circinator* (Gosse)	08 11 02 01
		clydona Harring & Myers	08 11 02 02
	Erignatha	*clastopis* (Gosse)	08 11 03 01
	Wierzejskiella	*elongata* (Glascott)	08 11 04 01
		ricciae (Harring)	08 11 04 02
	Encentrum	*eurycephalum* Wulfert	08 11 05 01
		felis (Muller)	08 11 05 02
		grande (Western)	08 11 05 03
		marinum (Dujardin)	08 11 05 04
		mustela (Milne)	08 11 05 05
		plicatum (Euferth)	08 11 05 06

		rousseleti (Lie-Pettersen)	08 11 05 07
		saundersiae (Hudson)	08 11 05 08
		voigti Wulfert	08 11 05 09
	Paradicranophorus	*hudsoni* (Glascott)	08 11 06 01
	Dicranophorus	*aspondus* Harring & Myers	08 11 07 01
		caudatus (Ehrenberg)	08 11 07 02
		forcipatus (Muller)	08 11 07 03
		grandis (Ehrenberg)	08 11 07 04
		hercules Wisniewski	08 11 07 05
		lutkeni (Bergendal)	08 11 07 06
		permollis (Gosse)	08 11 07 07
		rosa (Gosse)	08 11 07 08
		rostratus (Dizon-Nuttall&Freeman)	08 11 07 09
		uncinatus (Milne)	08 11 07 10
	Asplanchna	*brightwelli* Gosse	08 11 08 01
		girodi (Guerne)	08 11 08 02
		herricki Guerne	08 11 08 03
		intermedia Hudson	08 11 08 04
		priodonta Gosse	08 11 08 05
		sieboldi (Leydig)	08 11 08 06
	Harringia	*eupoda* (Gosse)	08 11 09 01
	Asplanchnopus	*multiceps* (Schrank)	08 11 10 01
Synchaetidae	*Polyarthra*	*delichoptera* Idelson	08 12 01 01
		euryptera Wierzejski	08 12 01 02
		longiremis Carlin	08 12 01 03
		major Burckhardt	08 12 01 04
		minor Voigt	08 12 01 05
		remata Skorikov	08 12 01 06
		vulgaris Carlin	08 12 01 07
	Synchaeta	*baltica* Ehrenberg	08 12 02 01
		bicornis Smith	08 12 02 02
		calva Ruttner-Kolisko	08 12 02 03
		cecilia Rousselet	08 12 02 04
		grandis Zacharias	08 12 02 05
		grimpei Remane	08 12 02 06
		gyrina Hood	08 12 02 07
		kitina Rousselet	08 12 02 08
		littoralis Rousselet	08 12 02 09
		oblonga Ehrenberg	08 12 02 10
		pectinata Ehrenberg	08 12 02 11
		stylata Wierzejski	08 12 02 12
		tavina Hood	08 12 02 13
		triophthalma Lauterbom	08 12 02 14
		vorax Rousselet	08 12 02 15
	Ploesoma	*hudsoni* (Imhof)	08 12 03 01
		lenticulare Herrick	08 12 03 02
		lynceus (Ehrenberg)	08 12 03 03
		truncatum (Levander)	08 12 03 04
Microcodinidae	*Microcodon*	*clavus* Ehrenberg	08 13 01 01
Testudinellidae	*Testudinella*	*caeca*	08 14 01 01
		clypeata (Muller)	08 14 01 02
		elliptica (Ehrenberg)	08 14 01 03
		emarginula (Stenroos)	08 14 01 04
		incisa (Tern)	08 14 01 05
		mucronata (Gosse)	08 14 01 06
		parva (Tern)	08 14 01 07
		patina (Herm)	08 14 01 08
		reflexa (Gosse)	08 14 01 09
		truncata (Gosse)	08 14 01 10
	Pompholyx	*complanata* Gosse	08 14 02 01
		sulcata Hudson	08 14 02 02

	Pedalia	mira (Hudson)	08 14 03 01
	Filinia	brachiata (Rousselet)	08 14 04 01
		cornuta (Weisse)	08 14 04 02
		longiseta (Ehrenberg)	08 14 04 03
		passa (Muller)	08 14 04 04
		terminalis (Plate)	08 14 04 05
Flosculariidae	Limnias	ceratophylli Schrank	08 15 01 01
		cornuella Rousselet	08 15 01 02
		melicerta Weisse	08 15 01 03
		myriophylli (Tatem)	08 15 01 04
	Floscularia	conifera (Hudson)	08 15 02 01
		janus (Hudson)	08 15 02 02
		melicerta (Ehrenberg)	08 15 02 03
		ringens (L.)	08 15 02 04
	Sinantherina	socialis (L.)	08 15 03 01
	Lacinularia	flosculosa (Muller)	08 15 04 01
	Beauchampia	crucigera (Dutrochet)	08 15 05 01
	Ptygura	beauchampia Edmondson	08 15 06 01
		brachiata (Hudson)	08 15 06 02
		brevis (Rousselet)	08 15 06 03
		cephaloceras Wright	08 15 06 04
		crystallina (Ehrenberg)	08 15 06 05
		furcillata (Kellicott)	08 15 06 06
		intermedia (Davis)	08 15 06 07
		lacunosa Wright	08 15 06 08
		longicornis (Davis)	08 15 06 09
		longipes Wills	08 15 06 10
		melicerta Ehrenberg	08 15 06 11
		mucicola (Kellicott)	08 15 06 12
		pilula (Cubitt)	08 15 06 13
		rotifer (Stenr)	08 15 06 14
		stygis (Gosse)	08 15 06 15
		tridorsicornis Wright	08 15 06 16
		velata (Gosse)	08 15 06 17
Conochilidae	Conochiloides	dossuarius (Hudson)	08 16 01 01
		natans (Seligo)	08 16 01 02
	Conochilus	hippocrepis (Schrank)	08 16 02 01
		unicornis Rousselet	08 16 02 02
Collothecidae	Stephanoceros	fimbriatus (Goldfuss)	08 17 01 01
	Collotheca	algicola (Hudson)	08 17 02 01
		ambigua (Hudson)	08 17 02 02
		annulata (Hood)	08 17 02 03
		calva (Hudson)	08 17 02 04
		campanulata (Dobie)	08 17 02 05
		coronetta (Cubitt)	08 17 02 06
		crateriformis Offord	08 17 02 07
		cucullata (Hood)	08 17 02 08
		cyclops (Cubitt)	08 17 02 09
		edentata (Collins)	08 17 02 10
		gossei (Hood)	08 17 02 11
		gracilipes Edmondson	08 17 02 12
		heptabrachiata (Schoch)	08 17 02 13
		hoodi (Hudson)	08 17 02 14
		libera (Zacharias)	08 17 02 15
		minuta Milne)	08 17 02 16
		mira (Hudson)	08 17 02 17
		moselii (Milne)	08 17 02 18
		mutabilis (Hudson)	08 17 02 19
		ornata (Ehrenberg)	08 17 02 20
		pelagica (Rousselet)	08 17 02 21

		quadrilobata (Hood)	08 17 02 22
		quadrinodosa Wright	08 17 02 23
		sessilis (Milne)	08 17 02 24
		spinata (Hood)	08 17 02 25
		staphanochaeta Edmondson	08 17 02 26
		tenuilobata (Anderson)	08 17 02 27
		torquilobata (Thorpe)	08 17 02 28
		trifidlobata (Pittock)	08 17 02 29
		trilobata (Collins)	08 17 02 30
	Cupelopagis	*vorax* (Leidy)	08 17 03 01
	Atrochus	*tentaculatus* Wierzejski	08 17 04 01
09 ACANTHOCEPHALA			09 00 00 00
10 NEMATODA	RHABDITIDA		
Cephalobidae	*Cephalobus*	*nanus* de Man	10 01 01 01
		persegnis Bastian	10 01 01 02
	Eucephalobus	*elongatus* de Man	10 01 02 01
		oxyuroides (de Man)	10 01 02 02
	Acrobeles	*ciliatus* Linstow	10 01 03 01
	Acrobebides	*emarginatus* (de Man)	10 01 04 01
Panagrolaimidae	*Panagrolaimus*	*rigidus* (Schneider)	10 02 01 01
		salinus Ever.	10 02 01 02
Teratocephalidae	*Teratocephalus*	*terrestris* (Butschli)	10 03 01 01
	Euteratocephalus	*crassidens* (de Man)	10 03 02 01
		palustris (de Man)	10 03 02 02
Rhabditidae	*Rhabditis*	*brevispina* (Claus)	10 04 01 01
		elongata (Schneider)	10 04 01 02
	Mesorhabditis	*monhystera* (Butschli)	10 04 02 01
	Diploscapter	*coronata* (Cobb)	10 04 03 01
Bunonematidae	*Bunonema*	*reticulatum* Richters	10 05 01 01
Diplogasteridae	*Diplogaster*	*rivalis* (Leydig)	10 06 01 01
	Diplogasteritus	*nudicapitatus* (Steiner)	10 06 02 01
	Goffartia	*variabilis* (Micoletzky)	10 06 03 01
	Mononchoides	*fictor* (Bastian)	10 06 04 01
		striatus (Butschli)	10 06 04 02
	TYLENCHIDA		
Tylenchidae	*Tylenchus*	*davainei* Bastian	10 07 01 01
		filiformis Butschli	10 07 01 02
		leptosoma de Man	10 07 01 03
	Aglenchus	*agricola* (de Man)	10 07 02 01
		bryophilus (Steiner)	10 07 02 02
	Ditylenchus	*intermedius* (de Man)	10 07 03 01
Hoplolaimidae	*Rotylenchus*	*robustus* (de Man)	10 08 01 01
	Hirschmannia	*gracilis* (de Man)	10 08 02 01
Criconematidae	*Criconemoides*	*rusticum* (Micoletzky)	10 09 01 01
Aphelenchoididae	*Aphelenchoides*	*helophilus* (de Man)	10 10 01 01
		parietinus (Bastian)	10 10 01 02
		saprophilus Frank	10 10 01 03
	AXONOLAIMIDA		
Plectidae	*Anaplectus*	*granulosus* (Bastian)	10 11 01 01
	Plectus	*cirratus* Bastian	10 11 02 01
		communis Butschli	10 11 02 02
		longicaudatus Butschli	10 11 02 03
		palustris de Man	10 11 02 04
		parietinus Bastian	10 11 02 05
		parvus Bastian	10 11 02 06

		rhizophilus de Man	10 11 02 07
		tenuis Bastian	10 11 02 08
	Wilsonema	auriculatum (Butschli)	10 11 03 01
Leptolaimidae	Rhabdolaimus	terrestris de Man	10 12 01 01
Camacolaimidae	Aphanolaimus	aquaticus Daday	10 13 01 01
		attentus de Man	10 13 01 02
Axonolaimidae	Cylindrolaimus	communis de Man	10 14 01 01

MONHYSTERIDA

Monhysteridae	Theristus	dubius (Butschli)	10 15 01 01
	Monhystera	dispar Bastian	10 15 02 01
		filiformis Bastian	10 15 02 02
		longicaudata Bastian	10 15 02 03
		paludicola de Man	10 15 02 04
		similis Butschli	10 15 02 05
		stagnalis Bastian	10 15 02 06
		vulgaris de Man	10 15 02 07

CHROMADORIDA

Cyatholaimidae	Achromadora	dubia (Butschli)	10 16 01 01
		ruricola (de Man)	10 16 01 02
		terricola (de Man)	10 16 01 03
	Ethmolaimus	pratensis de Man	10 16 02 01
Chromadoridae	Prochromadora	oerleyi (de Man)	10 17 01 01
	Chromadorina	bioculata (Schneider)	10 17 02 01
		viridis (Linstow)	10 17 02 02
	Punctodora	ratzeburgensis (Linstow)	10 17 03 01
	Chromadorita	leuckarti (de Man)	10 17 04 01

ENOPLIDA

Ironidae	Ironus	ignavus Bastian	10 18 01 01
		tenuicaudatus de Man	10 18 01 02
Tripylidae	Tripyla	filicaudata de Man	10 19 01 01
		glomerans Bastian	10 19 01 02
		monohystera de Man	10 19 01 03
		serifera de Man	10 19 01 04
	Trilobus	allophysis (Steiner)	10 19 02 01
		gracilis (Bastian)	10 19 02 02
		grandipapillatus (Brakenhoff)	10 19 02 03
		medius (Schneider)	10 19 02 04
		pellucidus (Bastian)	10 19 02 05
Onchulidae	Prismatolaimus	dolichurus de Man	10 20 01 01
		interdedius (Butschli)	10 20 01 02

DORYLAIMIDA

Mononchidae	Mononchus	papillatus Bastian	10 21 01 01
		truncatus Bastian	10 21 01 02
		tunbridgensis Bastian	10 21 01 03
	Prionchulus	muscorum (Dujardin)	10 21 02 01
Dorylaimidae	Prodorylaimus	longicaudatus (Butschli)	10 22 01 01
	Dorylaimus	filiformis Bastian	10 22 02 01
		flavomaculatus Linstow	10 22 02 02
		hofmaenneri Menzel	10 22 02 03
		saprophilus Peters	10 22 02 04
		stagnalis Dujardin	10 22 02 05
		tenuicaudatus Bastian	10 22 02 06
	Mesodorylaimus	bastiani (Butschli)	10 22 03 01
		mesonyctius (Kreis)	10 22 03 02
	Eudorylaimus	cartori (Bastian)	10 22 04 01
		iners (Bastian)	10 22 04 02
		intermedius (de Man)	10 22 04 03

		obtusicaudatus (Bastian)	10 22 04 04
	Paractinolaimus	macrolaimus (de Man)	10 22 05 01
Alaimidae	Alaimus	primitivus de Man	10 23 01 01
	Amphidelus	dolichurus (de Man)	10 23 02 01

11 GASTROTRICHA

Chaetonotidae	Chaetonotus	chuni Voigt	11 01 01 01
		cordiformis Grunspan	11 01 01 02
		disiuntus Grunspan	11 01 01 03
		hermaphroditus Remane	11 01 01 04
		hystrix Metschnikoff	11 01 01 05
		insigniformis Grunspan	11 01 01 06
		macrochaetus Zelinka	11 01 01 07
		maximus Ehrenberg	11 01 01 08
		persetosus Zelinka	11 01 01 09
		schultzei Metschnikoff	11 01 01 10
		similis Zelinka	11 01 01 11
		simrothi Voigt	11 01 01 12
		succinctus Voigt	11 01 01 13
		tabulatus Schmidt	11 01 01 14
		vorax Remane	11 01 01 15
		zelinkai Grunspan	11 01 01 16
	Lepidodermella	squamatum Dujardin	11 01 02 01
	Icthyolium	podura Metschnikoff	11 01 03 01
Dichaeturidae	Dichaetura	piscator Metschnikoff	11 02 01 01
Neogosseidae	Neogossea	antennigera Grunspan	11 03 01 01
Dasydytidae	Dasydytes	goniathrix Gosse	11 04 01 01
	Setopus	bisetosus Thompson	11 04 02 01

12 POLYZOA

GYMNOLAEMATA

| Paludicellidae | Paludicella | articulata (Ehrenberg) | 12 01 01 01 |

PHYLACTOLAEMATA

Fredericellidae	Fredericella	sultana (Blumenbach)	12 02 01 01
Plumatellidae	Plumatella	fruticosa Allman	12 03 01 01
		repens (L.)	12 03 01 02
		fungosa (Pallas)	12 03 01 03
		emarginata Allman	12 03 01 04
	Hyalinella	punctata (Hansock)	12 03 02 01
Lochopodidae	Lophopus	crystallinus (Pallas)	12 04 01 01
Cristatellidae	Cristatella	mucedo Cuvier	12 05 01 01

13 MOLLUSCA

GASTROPODA

Neritidae	Theodoxus	fluviatilis (L.)	13 01 01 01
Viviparidae	Viviparus	viviparus (L.)	13 02 01 01
		fasciatus (Muller)	13 02 01 02
Valvatidae	Valvata	cristata Muller	13 03 01 01
		macrostoma Morch	13 03 01 02
		piscinalis (Muller)	13 03 01 03
Hydrobiidae	Hydrobia	ventrosa (Montagu)	13 04 01 01
		ulvae (Pennant)	13 04 01 02
	Pseudamnicola	confusa (Frauenfeld)	13 04 02 01
	Potamopyrgus	jenkinsi (Smith)	13 04 03 01
	Bythinella	scholtzi (Schmidt)	13 04 04 01
	Bithynia	tentaculata (L.)	13 04 05 01
		leachi (Sheppard)	13 04 05 02
Assimineidae	Assiminea	grayana Fleming	13 05 01 01
Ellobiidae	Leucophytia	bidentata (Montagu)	13 06 01 01
	Phytia	myosotis (Draparnaud)	13 06 02 01

Lymnaeidae	Lymnaea	truncatula (Muller)	13 07 01 01
		glabra (Muller)	13 07 01 02
		palustris (Muller)	13 07 01 03
		catascopium Say	13 07 01 04
		stagnalis (L.)	13 07 01 05
		auricularia (L.)	13 07 01 06
		peregra (Muller)	13 07 01 07
	Myxas	glutinosa (Muller)	13 07 02 01
Physidae	Aplexa	hypnorum Fleming	13 08 01 01
	Physa	fontinalis (L.)	13 08 02 01
		heterostropha (Say)	13 08 02 02
Planorbidae	Planorbarius	corneus (L.)	13 09 01 01
	Menetus	dilatatus (Gould)	13 09 02 01
	Planorbis	carinatus Muller	13 09 03 01
		planorbis (L.)	13 09 03 02
		vorticulus Troschel	13 09 03 03
		vortex (L.)	13 09 03 04
		leucostoma Millet	13 09 03 05
		laevis Alder	13 09 03 06
		albus Muller	13 09 03 07
		acronicus Ferussac	13 09 03 08
		crista (L.)	13 09 03 09
		contortus (L.)	13 09 03 10
	Segmentina	complanata (L.)	13 09 04 01
		nitida (Muller)	13 09 04 02
Ancylidae	Acroloxus	lacustris (L.)	13 10 01 01
	Ancylus	fluviatilis Muller	13 10 02 01
Succineidae	Succinea	oblonga Draparnaud	13 11 01 01
		putris (L.)	13 11 01 02
		pfeifferi Rossmassler	13 11 01 03
		elegans Risso	13 11 01 04
Vertiginidae	Vertigo	antivertigo (Draparnaud)	13 12 01 01
		moulinsiana (Dupuy)	13 12 01 02
		lilljeborgi Westerlund	13 12 01 03
		angustior Jeffreys	13 12 01 04
Zonitidae	Zonitoides	nitidus (Muller)	13 13 01 01

14 BIVALVIA

Margaritiferidae	Margaritifera	margaritifera (L.)	14 01 01 01
Unionidae	Unio	pictorum (L.)	14 02 01 01
		tumidus Philipsson	14 02 01 02
	Anodonta	cygnaea (L.)	14 02 02 01
		anatina (L.)	14 02 02 02
		complanata Rossmassler	14 02 02 03
Sphaeriidae	Sphaerium	rivicola (Lamarck)	14 03 01 01
		corneum (L.)	14 03 01 02
		transversum (Say)	14 03 01 03
		lacustre (Muller)	14 03 01 04
	Pisidium	amnicum (Muller)	14 03 02 01
		casertanum (Poli)	14 03 02 02
		conventus Clessin	14 03 02 03
		personatum Malm	14 03 02 04
		obtusale (Lamarck)	14 03 02 05
		milium Held	14 03 02 06
		pseudosphaerium (Favre)	14 03 02 07
		subtruncatum Malm	14 03 02 08
		supinum Schmidt	14 03 02 09
		henslowanum (Sheppard)	14 03 02 10

		lilljeborgii (Clessin)	14 03 02 11
		hibernicum Westerlund	14 03 02 12
		nitidum Jenyns	14 03 02 13
		pulchellum Jenyns	14 03 02 14
		moitessierianum Paladilhe	14 03 02 15
		tenuilineatum Stelfox	14 03 02 16
Dreissenidae	*Dreissena*	*polymorpha* (Pallas)	14 04 01 01

15 ANNELIDA   POLYCHAETA

| Nereidae | *Nereis* | *diversicolor* Muller | 15 01 01 01 |

16   OLIGOCHAETA

Aeolosomatidae	*Aeolosoma*	*quaternarium* Ehrenberg	16 01 01 01
		hemprici Ehrenberg	16 01 01 02
		headleyi Beddard	16 01 01 03
		tenebrarum Vejdovsky	16 01 01 04
		variegatum Vejdovsky	16 01 01 05
		beddardi Michaelsen	16 01 01 06
Naididae	*Chaetogaster*	*diastrophus* (Gruithuisen)	16 02 01 01
		langi Bretscher	16 02 01 02
		diaphanus (Gruithuisen)	16 02 01 03
		cristallinus Vejdovsky	16 02 01 04
		limnaei von Baer	16 02 01 05
	Paranais	*litoralis* (Muller)	16 02 02 01
	Homochaeta	*naidina* Bretscher	16 02 03 01
	Specaria	*josinae* (Vejdovsky)	16 02 04 01
	Uncinais	*uncinata* (Orsted)	16 02 05 01
	Ophidonais	*serpentina* (Muller)	16 02 06 01
		reckei Floericke	16 02 06 02
	Nais	*communis* Piguet	16 02 07 01
		variabilis Piguet	16 02 07 02
		simplex Piguet	16 02 07 03
		alpina Sperber	16 02 07 04
		barbata Muller	16 02 07 05
		pseudobtusa Piguet	16 02 07 06
		elinguis Muller	16 02 07 07
		heterochaeta Benham	16 02 07 08
	Slavina	*appendiculata* (d'Udekem)	16 02 08 01
	Vejdovskyella	*comata* (Vejdovsky)	16 02 09 01
		intermedia (Bretscher)	16 02 09 02
	Arcteonais	*lomondi* (Martin)	16 02 10 01
	Ripistes	*parasita* (Schmidt)	16 02 11 01
	Stylaria	*lacustris* (L.)	16 02 12 01
	Piguetiella	*blanci* (Piguet)	16 02 13 01
	Dero	*digitata* (Muller)	16 02 14 01
		obtusa d'Udekem	16 02 14 02
		perrieri Bousfield	16 02 14 03
		latissima Bousfield	16 02 14 04
		roseola Nicholls	16 02 14 05
		furcata Oken	16 02 14 06
	Aulophorus	*furcatus* (Muller)	16 02 15 01
	Pristina	*menoni* (Aiyer)	16 02 16 01
		idrensis Sperber	16 02 16 02
		foreli (Piguet)	16 02 16 03
		aequiseta Bourne	16 02 16 04
		longiseta Ehrenberg	16 02 16 05
Tubificidae	*Tubifex*	*tubifex* (Muller)	16 03 01 01
		ignota (Stolc)	16 03 01 02

		ignotus (Stolc)	16 03 01 03
		nerthus Michaelsen	16 03 01 04
		costatus Claparede	16 03 01 05
		pseudogaster (Dahl)	16 03 01 06
		newaensis (Michaelsen)	16 03 01 07
	Psammoryctes	*barbata* (Grube)	16 03 02 01
		albicola (Michaelsen)	16 03 02 02
	Limnodrilus	*claparedeanus* Ratzel	16 03 03 01
		hoffmeisteri Claparede	16 03 03 02
		udekemianus Claparede	16 03 03 03
		parvus Southern	16 03 03 04
		profundicola (Verrill)	16 03 03 05
	Peloscolex	*ferox* (Eisen)	16 03 04 01
		benedeni (d'Udekem)	16 03 04 02
		velutinus (Grube)	16 03 04 03
		specrosus (Hrabe)	16 03 04 04
	Potamothrix	*hammoniensis* (Michaelsen)	16 03 05 01
		bavaricus (Oschmann)	16 03 05 02
		moldaviensis (Vejdovsky&Mrazek)	16 03 05 03
	Rhyacodrilus	*coccineus* (Vejdovsky)	16 03 06 01
		falciformis Bretscher	16 03 06 02
	Bothrioneurum	*vejdovskyanum* Stolc	16 03 07 01
	Ilyodrilus	*templetoni* Southern	16 03 08 01
	Branchiura	*sowerbyi* Beddard	16 03 09 01
	Monopylephorus	*rubroniveus* Levinsen	16 03 10 01
		irrorata (Verrill)	16 03 10 02
		parvus Ditlevsen	16 03 10 03
	Aulodrilus	*pluriseta* Piguet	16 03 11 01
	Clitellio	*arenarius* (Muller)	16 03 12 01
Enchytraeidae	*Mesenchytraeus*	*sanguineus* Nielsen&Christensen	16 04 01 01
		armatus Levinsen	16 04 01 02
	Cernosvitoviella	*atrata* (Bretscher)	16 04 02 01
		immota (Knollner)	16 04 02 02
	Cognettia	*sphagnetorum* (Vedjovsky)	16 04 03 01
		glandulosa (Michaelsen)	16 04 03 02
		cognetti (Issel)	16 04 03 03
	Henlea	*perpusilla* Friend	16 04 04 01
		ventriculosa (d'Udekem)	16 04 04 02
	Bucholzia	*fallax* Michaelsen	16 04 05 01
	Fridericia	*bisetosa* (Levinsen)	16 04 06 01
		leydigi (Vedjovsky)	16 04 06 02
		perrieri (Vedjovsky)	16 04 06 03
		galba (Hoffmeister)	16 04 06 04
		hegemon (Vedjovsky)	16 04 06 05
		aurita Issel	16 04 06 06
	Enchytraeus	*albidus* Henle	16 04 07 01
		buchholzi Vedjovsky	16 04 07 02
		minutus Nielsen & Christensen	16 04 07 03
	Lumbricillus	*rivalis* Levinsen	16 04 08 01
		lineatus (Muller)	16 04 08 02
		pagenstecheri (Ratzel)	16 04 08 03
		arenarius (Michaelsen)	16 04 08 04
	Marionina	*spicula* (Leuckart)	16 04 09 01
		appendiculata Nielsen&Christensen	16 04 09 02
Haplotaxidae	*Haplotaxis*	*gordioides* (Hartmann)	16 05 01 01
Lumbriculidae	*Lumbriculus*	*variegatus* (Muller)	16 06 01 01
	Stylodrilus	*heringianus* Claparede	16 06 02 01

	Eclipidrilus	*lacustris* (Verill)	16 06 03 01
	Trichodrilus	*hrabei* Cook	16 06 04 01
		cantabrigiensis (Beddard)	16 06 04 02
		icenorum Beddard	16 06 04 03
	Rhynchelmis	*limosella* Hoffmeister	16 06 05 01
Glossoscolecidae	*Sparganophilus*	*tamesis* Benham	16 07 01 01
Lumbricidae	*Allolobophora*	*chlorotica* (Savigny)	16 08 01 01
		rosea (Savigny)	16 08 01 02
	Dendrobaena	*rubida* (Savigny)	16 08 02 01
	Eiseniella	*tatraedra* (Savigny)	16 08 03 01
	Helodrilus	*oculatus* Hoffmeister	16 08 04 01
	Octoclasion	*cyaneum* (Savigny)	16 08 05 01
		lacteum (Oerley)	16 08 05 02
Dorydrilidae	*Dorydrilus*	*michaelseni* Piguet	16 09 01 01
Branchiobdellidae	*Branchiobdella*	*astaci* Odier	16 10 01 01

17   HIRUDINEA

Piscicolidae	*Piscicola*	*geometra* (L.)	17 01 01 01
Glossiphoniidae	*Theromyzon*	*tessulatum* (Muller)	17 02 01 01
	Hemiclepsis	*marginata* (Muller)	17 02 02 01
	Glossiphonia	*heteroclita* (L.)	17 02 03 01
		complanata (L.)	17 02 03 02
	Batracobdella	*paludosa* (Carena)	17 02 04 01
	Helobdella	*stagnalis* (L.)	17 02 05 01
Hirudidae	*Haemopsis*	*sanguisuga* (L.)	17 03 01 01
	Hirudo	*medicinalis* L.	17 03 02 01
Erpobdellidae	*Erpobdella*	*testacea* (Savigny)	17 04 01 01
		octoculata (L.)	17 04 01 02
	Dina	*lineata* (Muller)	17 04 02 01
	Trocheta	*subviridis* Dutrochet	17 04 03 01
		bykowskii Gedroyc	17 04 03 02

18   ARTHROPODA: OTHERS   TARDIGRADA

Scutechiniscidae	*Echiniscus*	*reticulatus* Murray	18 01 01 01
		gladiator Murray	18 01 01 02
		wendti Richters	18 01 01 03
		meticulatus Murray	18 01 01 04
		oihonnae Richters	18 01 01 05
		granulatus Doyere	18 01 01 06
		spitzbergensis Scourfield	18 01 01 07
		quadrispinosus Richters	18 01 01 08
		tympanista Murray	18 01 01 09
Arctiscidae	*Milnesium*	*tardigradum* Doyers	18 02 01 01
Macrobiotidae	*Macrobiotus*	*hufelandii* Schultze	18 03 01 01
		echinogenitus Richters	18 03 01 02
		dispar Murray	18 03 01 03
		pullari Murray	18 03 01 04
		hastatus Murray	18 03 01 05
		ambiguus Murray	18 03 01 06
		macronyx Dujardin	18 03 01 07
	Calohypsibuis	*ornatus* (Richters)	18 03 02 01
	Isohypsibius	*tuberculatus* (Plate)	18 03 03 01
		sattleri (Richters)	18 03 03 02
		papillifer (Murray)	18 03 03 03
		annulatus (Murray)	18 03 03 04
		augusti (Murray)	18 03 03 05

		convergens (Urban)	18 03 03 06
		prosostomus (Thulim)	18 03 03 07
		schaudinni (Richters)	18 03 03 08
	Hypsibius	*oberhauseri* (Doyere)	18 03 04 01
		dujardini (Doyere)	18 03 04 02
		arcticus (Murray)	18 03 04 03
	Diphascon	*chilenensis* (Plate)	18 03 05 01
		spitzbergensis (Richters)	18 03 05 02
		angustatus (Murray)	18 03 05 03
		scoticus (Murray)	18 03 05 04
		bullatus (Murray)	18 03 05 05
		oculatus (Murray)	18 03 05 06

19 HYDRACARINA

Family	Genus	Species	Code
Hydrovolziidae	*Hydrovolzia*	*placophora* (Monti)	19 01 01 01
Hydrachnidae	*Hydrachna*	*cruenta* Muller	19 02 01 01
		skorikowi Piersig	19 02 01 02
	Diplohydrachna	*conjecta* Koenike	19 02 02 01
		distincta Koenike	19 02 02 02
		georgei Soar	19 02 02 03
		globosa (Geer)	19 02 02 04
	Rhabdohydrachna	*bivirgulata* Piersig	19 02 03 01
		comosa Koenike	19 02 03 02
		geographica Muller	19 02 03 02
		halberti Soar	19 02 03 04
		incisa Halbert	19 02 03 05
		leegei Koenike	19 02 03 06
		levis Williamson	19 02 03 07
		processifera Koenike	19 02 03 08
		williamsoni Soar	19 02 03 09
	(Hydrachna)	*ferox* Dalyell	19 02 04 01
		placida Dalyell	19 02 04 02
		punctata Dalyell	19 02 04 03
		sparsa Dalyell	19 02 04 04
		spinifera Dalyell	19 02 04 05
		varia Dalyell	19 02 04 06
Limnocharidae	*Limnochares*	*aquatica* (L.)	19 03 01 01
Elayidae	*Eylais*	*bicornuta* Halbert	19 04 01 01
		bisinuosa Piersig	19 04 01 02
		celtica Halbert	19 04 01 03
		cocinea (Shaw)	19 04 01 04
		discreta Koenike	19 04 01 05
		dividuus Soar	19 04 01 06
		extendens (Muller)	19 04 01 07
		gigas Piersig	19 04 01 08
		hamata Koenike	19 04 01 09
		infundibulifera Koenike	19 04 01 10
		insularis Thor	19 04 01 11
		koenikei Halbert	19 04 01 12
		meridionalis Thon	19 04 01 13
		mulleri Koenike	19 04 01 14
		neglecta Thor	19 04 01 15
		relicta Halbert	19 04 01 16
		rimosa Piersig	19 04 01 17
		symmetrica Halbert	19 04 01 18
		tantilla Koenike	19 04 01 19
		wilsoni Soar	19 04 01 20
Piersigiidae	*Piersigia*	*intermedia* Williamson	19 05 01 01
		koenikei Viets	19 05 01 02
Hydryphantidae	*Hydryphantes*	*bayeri* Piscarovic	19 06 01 01
		crassipalpis Koenike	19 06 01 02
		dispar (Schaub)	19 06 01 03

		frici Thon	19 06 01 04
		placationis Thon	19 06 01 05
		ruber (Geer)	19 06 01 06
	Polyhydryphantes	*flexuosus* (Koenike)	19 06 02 01
	Lundbladia	*petrophila* (Michael)	19 06 03 01
	Panisellus	*thienemanni* (Viets)	19 06 04 01
	Vietsia	*scutata* (Protz)	19 06 05 01
	Thyopsis	*cancellata* (Protz)	19 06 06 01
	Panisus	*michaeli* Koenike	19 06 07 01
		torrenticolus Piersig	19 06 07 02
	Panisopsis	*vigilans* (Piersig)	19 06 08 01
	Thyasella	*mandibularis* (Lundblad)	19 06 09 01
	Thyas	*barbigera* Viets	19 06 10 01
		dirempta Koenike	19 06 10 02
		extendens George	19 06 10 03
		pachystoma Koenike	19 06 10 04
		rivalis Koenike	19 06 10 05
	Zschokkea	*oblonga* (Koenike)	19 06 11 01
	Parathyas	*thoracata* (Piersig)	19 06 12 01
	Euthyas	*truncata* (Neuman)	19 06 13 01
	Protzia	*eximia* (Protz)	19 06 14 01
	Calonyx	*rotunda* Walter	19 06 15 01
	Wandesia	*racovitzai* Gledhill	19 06 16 01
	Pseudhydryphantes	*parvulus* Viets	19 06 17 01
Hydrodromidae	*Hydrodroma*	*despicuens* (Muller)	19 07 01 01
Sperchonidae	*Sperchonopsis*	*verrucosa* (Protz)	19 08 01 01
	Sperchon	*brevirostris* Koenike	19 08 02 01
		clupeifer Piersig	19 08 02 02
		denticulatus Koenike	19 08 02 03
		glandulosus Koenike	19 08 02 04
		hibernicus Halbert	19 08 02 05
		hispidus Koenike	19 08 02 06
		longirostris Koenike	19 08 02 07
		longissimus Viets	19 08 02 08
		ornatus Halbert	19 08 02 09
		papillosus Thor	19 08 02 10
		setiger Thor	19 08 02 11
		squamosus Kramer	19 08 02 12
		thienemanni Koenike	19 08 02 13
Teutoniidae	*Teutonia*	*cometes* (Koch)	19 09 01 01
Anisitsiellidae	*Bandakia*	*concreta* Thor	19 10 01 01
	Dartia	*harrisi* (Soar)	19 10 02 01
Lebertiidae	*Lebertia*	*africana* Walter	19 11 01 01
		areolata Halbert	19 11 01 02
		brunnea Halbert	19 11 01 03
		castalia Viets	19 11 01 04
		cognata Koenike	19 11 01 05
		compacta Halbert	19 11 01 06
		dalmatica Viets	19 11 01 07
		fimbriata Thor	19 11 01 08
		flumenia Halbert	19 11 01 09
		hirtipalpis Halbert	19 11 01 10
		laticoxalis Viets	19 11 01 11
		maglioi Thor	19 11 01 12
		minuta Halbert	19 11 01 13
		rufipes Koenike	19 11 01 14
		sparsicapillata Thor	19 11 01 15

		tenuipalpis Halbert	19 11 01 16
		tenuistriata Viets	19 11 01 17
	Pilolebertia	*crassipalpis* Halbert	19 11 02 01
		curvipalpis Halbert	19 11 02 02
		hibernica Viets	19 11 02 03
		inaequalis (Koch)	19 11 02 04
		insignis Neuman	19 11 02 05
		minuticornis Viets	19 11 02 06
		pachypalpis Sokolow	19 11 02 07
		plauta Halbert	19 11 02 08
		porosa Thor	19 11 02 09
		rotunda Halbert	19 11 02 10
		valenciana Viets	19 11 02 11
	Pseudolebertia	*glabra* Thor	19 11 03 01
		mollis Halbert	19 11 03 02
		salebrosa Koenike	19 11 03 03
	Hexalebertia	*dubia* Thor	19 11 04 01
		novipalpis Halbert	19 11 04 02
		sefvei Walter	19 11 04 03
		stigmatifera Thor	19 11 04 04
	Mixolebertia	*densa* Koenike	19 11 05 01
		halberti Koenike	19 11 05 02
		oudemansi Koenike	19 11 05 03
	Frontipoda	*carpenteri* (Halbert)	19 11 06 01
		musculus (Muller)	19 11 06 02
	Oxus	*nodigerus* Koenike	19 11 07 01
		ovalis (Muller)	19 11 07 02
		strigatus (Muller)	19 11 07 03
	Gnaphiscus	*setosus* (Koenike)	19 11 08 01
Torrenticolidae	*Torrenticola*	*amplexa* (Koenike)	19 12 01 01
		andrei (Angelier)	19 12 01 02
		anomala (Koch)	19 12 01 03
		brevirostris (Halbert)	19 12 01 04
		elliptica Maglio	19 12 01 05
		halberti (Thor)	19 12 01 06
		maglioi (Koenike)	19 12 01 07
		parvipalpus (Halbert)	19 12 01 08
		robusta (Halbert)	19 12 01 09
		thori (Halbert)	19 12 01 10
	Monatractides	*madritensis* (Viets)	19 12 02 01
Limnesiidae	*Limnesia*	*connata* Koenike	19 13 01 01
		fulgida Koch	19 13 01 02
		koenikei Piersig	19 13 01 03
		maculata (Muller)	19 13 01 04
		undulata (Muller)	19 13 01 05
Hygrobatidae	*Hygrobates*	*calliger* Piersig	19 14 01 01
		fluviatilis (Strom)	19 14 01 02
		foreli (Lebert)	19 14 01 03
		longipalpis (Hermann)	19 14 01 04
		longiporus Thor	19 14 01 05
		nigromaculatus Lebert	19 14 01 06
		trigonicus Koenike	19 14 01 07
	Rivobates	*norvegicus* Thor	19 14 02 01
	Atractides	*denticulatus* (Walter)	19 14 03 01
		elongatus (Halbert)	19 14 03 02
		gibberipalpis Piersig	19 14 03 03
		latipalpis (Motas & Tanasachi)	19 14 03 04
		longipes (Halbert)	19 14 03 05
		magnirostris (Motas & Tanasachi)	19 14 03 06
		nodipalpis (Thor)	19 14 03 07
		pachydermis (Halbert)	19 14 03 08

		pavesii Maglio	19 14 03 09
		spinipes Koch	19 14 03 10
		tener (Thor)	19 14 03 11
	Octomegapus	*octoporus* Piersig	19 14 04 01
	Unionicola	*crassipes* (Muller)	19 14 05 01
		gracilipalpis (Viets)	19 14 05 02
	Pentatax	*aculeata* (Koenike)	19 14 06 01
		bonzi (Claparede)	19 14 06 02
		figuralis (Koch)	19 14 06 03
		intermedia (Koenike)	19 14 06 04
	Parasitatax	*ypsilophora* (Bonz)	19 14 07 01
	Neumania	*callosa* (Koenike)	19 14 08 01
		deltoides (Piersig)	19 14 08 02
		limosa (Koch)	19 14 08 03
		spinipes (Muller)	19 14 08 04
		vernalis (Muller)	19 14 08 05
	Soarella	*papillosa* (Soar)	19 14 09 01
Feltriidae	*Feltria*	*cornuta* Walter	19 15 01 01
		denticulata Angelier	19 15 01 02
		minuta Koenike	19 15 01 03
		rouxi Walter	19 15 01 04
		subterranea Viets	19 15 01 05
	Azugofeltria	*motasi* (Schwoerbel)	19 15 02 01
Pionidae	*Huitfeldtia*	*rectipes* Thor	19 16 01 01
	Piona	*affinis* (Koch)	19 16 02 01
		alata (Thor)	19 16 02 02
		alpicola (Neuman)	19 16 02 03
		ambigua Piersig	19 16 02 04
		carnea (Koch)	19 16 02 05
		clavicornis (Muller)	19 16 02 06
		coccinea (Koch)	19 16 02 07
		conglobata (Koch)	19 16 02 08
		discrepans (Koenike)	19 16 02 09
		disparilis (Koenike)	19 16 02 10
		fallax (Thon)	19 16 02 11
		longipalpis (Krendowsky)	19 16 02 12
		neumani (Koenike)	19 16 02 13
		nodata (Muller)	19 16 02 14
		obturbans (Piersig)	19 16 02 15
		paucipora (Thor)	19 16 02 16
		pusilla (Neuman)	19 16 02 17
		tuberifera Viets	19 16 02 18
		variabilis (Koch)	19 16 02 19
	Nautarachna	*crassa* (Koenike)	19 16 03 01
	Pionella	*karamani* (Viets)	19 16 04 01
	Wettina	*podagrica* (Koch)	19 16 05 01
	Hydrochoreutes	*krameri* Piersig	19 16 06 01
		ungulatus (Koch)	19 16 06 02
	Tiphys	*bullatus* (Thor)	19 16 07 01
		lapponicus (Neuman)	19 16 07 02
		latipes (Muller)	19 16 07 03
		ornatus Koch	19 16 07 04
		scaurus (Koenike)	19 16 07 05
		torris (Muller)	19 16 07 06
	Pionides	*ensifer* (Koenike)	19 16 08 01
	Aceropsis	*pistillifer* (Koenike)	19 16 09 01
	Pionopsis	*lutescens* (Hermann)	19 16 10 01
	Pionacercus	*leuckarti* Piersig	19 16 11 01
		norvegicus Thor	19 16 11 02

		pyriformis Soar	19 16 11 03
		uncinatus (Koenike)	19 16 11 04
	Pionacercopsis	*vatrax* (Koch)	19 16 12 01
	Forelia	*brevipes* (Neuman)	19 16 13 01
		liliacea Muller	19 16 13 02
		variegator (Koch)	19 16 13 03
	Pseudofeltria	*scourfieldi* Soar	19 16 14 01
Aturidae	*Barbaxonella*	*angulata* (Viets)	19 17 01 01
	Vietsaxona	*lundbladi* Motas & Tanasachi	19 17 02 01
	Axonopsis	*complanata* (Muller)	19 17 03 01
	Hexaxonopsis	*romijni* Viets	19 17 04 01
	Brachypoda	*versicolor* (Muller)	19 17 05 01
	Ocybrachypoda	*celeripes* Viets	19 17 06 01
	Ljania	*bipapillata* Thor	19 17 07 01
	Lethaxona	*cavifrons* Szalay	19 17 08 01
	Aturus	*brachypus* Viets	19 17 09 01
		crinitus Thor	19 17 09 02
		intermedius Protz	19 17 09 03
		scaber Kramer	19 17 09 04
	Kongsbergia	*dypeata* Szalay	19 17 10 01
		largaiollii (Maglio)	19 17 10 02
		materna Thor	19 17 10 03
		vietsi Halbert	19 17 10 04
Mideidae	*Midea*	*orbiculata* (Muller)	19 18 01 01
Momoniidae	*Momonia*	*falcipalpis* Halbert	19 19 01 01
	Stygomomonia	*latipes* Szalay	19 19 02 01
Mideopsidae	*Mideopsis*	*crassipes* Soar	19 20 01 01
		orbicularis (Muller)	19 20 01 02
	Xystonotus	*willmani* (Viets)	19 20 02 01
Neoacaridae	*Neoacarus*	*hibernicus* Halbert	19 21 01 01
Athienemanniidae	*Chelomideopsis*	*annemiae* Romijn	19 22 01 01
	Mundamella	*germanica* Viets	19 22 02 01
Hungarohydracaridae	*Hungarohydracarus*	*subterraneus* Szalay	19 23 01 01
Arrenuridae	*Arrenurus*	*abbreviator* Berlese	19 24 01 01
		affinis Koenike	19 24 01 02
		albator (Muller)	19 24 01 03
		batillifer Koenike	19 24 01 04
		bicuspidator Berkse	19 24 01 05
		bruzelii Koenike	19 24 01 06
		claviger Koenike	19 24 01 07
		compactus Piersig	19 24 01 08
		crassicaudatus Kramer	19 24 01 09
		crenatus Koenike	19 24 01 10
		cuspidifer Piersig	19 24 01 11
		fimbriatus Koenike	19 24 01 12
		latus Barrois & Moniez	19 24 01 13
		leuckarti Piersig	19 24 01 14
		maculator (Muller)	19 24 01 15
		neumani Piersig	19 24 01 16
		nobilis Neuman	19 24 01 17
		ornatus George	19 24 01 18
		robustus Koenike	19 24 01 19
		tricuspidator (Muller)	19 24 01 20
		virens Neuman	19 24 01 21

	Megaluracarus	*adnatus* Koenike	19 24 02 01
		buccinator (Muller)	19 24 02 02
		curtus George	19 24 02 03
		cylindratus Piersig	19 24 02 04
		eugeminus Piersig	19 24 02 05
		globator (Muller)	19 24 02 06
		insperatus George	19 24 02 07
		mediorotundatus Thor	19 24 02 08
		membranator Thor	19 24 02 09
		mulleri Koenike	19 24 02 10
		pyriformis George	19 24 02 11
		scourfieldi Soar	19 24 02 12
		securiformis Piersig	19 24 02 13
		soari George	19 24 02 14
		zachariae Koenike	19 24 02 15
	Truncaturus	*fontinalis* Viets	19 24 03 01
		knauthei Koenike	19 24 03 02
		nodosus Koenike	19 24 03 03
		stecki Koenike	19 24 03 04
		truncatellus (Muller)	19 24 03 05
	Micruracarus	*bifidicodulus* Piersig	19 24 04 01
		bipapillosus Halbert	19 24 04 02
		biscissus Lebert	19 24 04 03
		britannorum Viets	19 24 04 04
		forpicatus Neuman	19 24 04 05
		inexploratus Viets	19 24 04 06
		integrator (Muller)	19 24 04 07
		longiusculus George	19 24 04 08
		novus George	19 24 04 09
		octagonus Halbert	19 24 04 10
		perforatus George	19 24 04 11
		sculptus Halbert	19 24 04 12
		sinuator (Muller)	19 24 04 13
	(Arrenurus)	*luteus* George	19 24 05 01
		mollis George	19 24 05 02
Limnohalacaridae	*Porochalacarus*	*alpinus* (Thor)	19 25 01 01
	Lobohalacarus	*dolgarae* Green	19 25 02 01
		weberi (Ronijn & Viets)	19 25 02 02
	Limnohalacarus	*wackeri* (Walter)	19 25 03 01
	Soldanellonyx	*chappuisi* Walter	19 25 04 01
		monardi Walter	19 25 04 02
	Parasoldanellonyx	*parviscutatus* (Walter)	19 25 05 01
	Porolohmannella	*violacea* (Kramer)	19 25 06 02

20 ORIBATEI

Astegistidae	*Astegistes*	*pilosus* (Koch)	20 01 01 01
Hydrozetidae	*Hydrozetes*	*lacustris* (Michael)	20 02 01 01
		lemnae (Coggi)	20 02 01 02
Ceratozetidae	*Ceratozetes*	*furcatus* (Pearc & Willman)	20 03 01 01

21 ARANEAE

Agelinidae	*Argyroneta*	*aquatica* (Clerck)	21 01 01 01

22 ARTHROPODA: CRUSTACEA ANOSTRACA

Artemiidae	*Artemia*	*salina* L.	22 01 01 01
Chirocephalidae	*Chirocephalus*	*diaphanus* Prevost	22 02 01 01

23 NOTOSTRACA

Triopsidae	*Triops*	*cancriformis* L.	23 01 01 01

24 CLADOCERA

Sididae	*Diaphanosoma*	*brachyurum* (Lieven)	24 01 01 01
		leuchtenbergianum Fischer	24 01 01 02

	Latona	*setifera* (Muller)	24 01 02 01
	Sida	*crystallina* (Muller)	24 01 03 01
Holopedidae	*Holopedium*	*gibberum* Zaddach	24 02 01 01
Daphniidae	*Ceriodaphnia*	*dubia* Richard	24 03 01 01
		laticaudata Muller	24 03 01 02
		megops Sars	24 03 01 03
		pulchella Sars	24 03 01 04
		quadrangula (Muller)	24 03 01 05
		reticulata (Jurine)	24 03 01 06
		setosa Matile	24 03 01 07
	Daphnia	*ambigua* Scourfield	24 03 02 01
		atkinsoni Baird	24 03 02 02
		cucullata Sars	24 03 02 03
		curvirostris Eylmann	24 03 02 04
		hyalina Leydig	24 03 02 05
		galeata Sars	24 03 02 06
		longispina (Muller)	24 03 02 07
		magna Straus	24 03 02 08
		obtusa Kurz	24 03 02 09
		pulex (De Geer)	24 03 02 10
	Moina	*brachiata* (Jurine)	24 03 03 01
		macrocopa (Straus)	24 03 03 02
	Scapholeberis	*aurita* (Fischer)	24 03 04 01
		mucronata (Muller)	24 03 04 02
	Simocephalus	*exspinosus* (Koch)	24 03 05 01
		serrulatus (Koch)	24 03 05 02
		vetulus (Muller)	24 03 05 03
	Bosmina	*longirostris* (Muller)	24 03 06 01
		coregoni Baird	24 03 06 02
Macrothricidae	*Acantholeberis*	*curvirostris* (Muller)	24 04 01 01
	Drepanothrix	*dentata* (Euren)	24 04 02 01
	Ilyocryptus	*acutifrons* Sars	24 04 03 01
		agilis Kurz	24 04 03 02
		sordidus (Lieven)	24 04 03 03
	Lathonura	*rectirostris* (Muller)	24 04 04 01
	Macrothrix	*hirsuticornis* Norman & Brady	24 04 05 01
		laticornis (Jurine)	24 04 05 02
		rosea (Jurine)	24 04 05 03
	Ophryoxus	*gracilis* Sars	24 04 06 01
	Streblocerus	*serricaudatus* (Fischer)	24 04 07 01
Chydoridae	*Acroperus*	*harpae* Baird	24 05 01 01
		elongata (Sars)	24 05 01 02
	Alona	*affinis* (Leydig)	24 05 02 01
		costata Sars	24 05 02 02
		guttata Sars	24 05 02 03
		intermedia (Sars)	24 05 02 04
		protzi Hartwig	24 05 02 05
		quadrangularis (Muller)	24 05 02 06
		rectangula Sars	24 05 02 07
		rustica Scott	24 05 02 08
		weltneri Keilhack	24 05 02 09
		elegans Kurz	24 05 02 10
	Alonella	*excisa* (Fischer)	24 05 03 01
		exigua (Lilljeborg)	24 05 03 02
		nana (Baird)	24 05 03 03
	Disparalona	*rostrata* (Koch)	24 05 04 01
	Tretocephala	*ambigua* (Lilljeborg)	24 05 05 01
	Anchistropus	*emarginatus* Sars	24 05 06 01
	Camptocercus	*lilljeborgi* Schodler	24 05 07 01

		rectirostris Schodler	24 05 07 02
Chydorus		gibbus Lilljeborg	24 05 08 01
		latus Sars	24 05 08 02
		ovalis Kurz	24 05 08 03
		piger Sars	24 05 08 04
		sphaericus (Muller)	24 05 08 05
Pseudochydorus		globosus (Baird)	24 05 09 01
Eurycercus		glacialis Lilljeborg	24 05 10 01
		lamellatus Muller	24 05 10 02
Graptoleberis		testudinaria (Fischer)	24 05 11 01
Kurzia		latissima (Kurz)	24 05 12 01
Leydigia		acanthocercoides (Fischer)	24 05 13 01
		leydigii (Schodler)	24 05 13 02
Monospilus		dispar Sars	24 05 14 01
Oxyurella		tenuicaudis (Sars)	24 05 15 01
Peracantha		truncata (Muller)	24 05 16 01
Pleuroxua		aduncus (Jurine)	24 05 17 01
		laevis Sars	24 05 17 02
		trigonellus (Muller)	24 05 17 03
		uncinatus Baird	24 05 17 04
		denticulatus Birge	24 05 17 05
Dunhevedia		crassa King	24 05 18 01
Rhynchotalona		falcata (Sars)	24 05 19 01

Polyphemidae	Polyphemus	pediculus L.	24 06 01 01
	Bythotrephes	longimanus Leydig	24 06 02 01
		cederstroni Schodler	24 06 02 02
Leptodoridae	Leptodora	kindti (Focke)	24 07 01 01

25 OSTRACODA

Cyprididae	Candona	angulata Muller	25 01 01 01
		candida (Muller)	25 01 01 02
		neglecta Sars	25 01 01 03
		lobipes Hartwig	25 01 01 04
		marchica Hartwig	25 01 01 05
		rostrata Brady & Norman	25 01 01 06
		sarsi Hartwig	25 01 01 07
		insculpta Muller	25 01 01 08
		parallela Muller	25 01 01 09
		fabaeformis Fischer	25 01 01 10
		lapponica Ekman	25 01 01 11
		acuminata Fischer	25 01 01 12
		caudata Kaufmann	25 01 01 13
		hyalina Brady & Robertson	25 01 01 14
		protzi Hartwig	25 01 01 15
		vavrai Kaufmann	25 01 01 16
		nitens Robertson	25 01 01 17
		reducta Alm	25 01 01 18
		wedgewoodii Lowndes	25 01 01 19
	Pseudocandona	elongata Holmes	25 01 02 01
	Candonopsis	kingsleii Brady	25 01 03 01
	Paracandona	euplectella (Brady & Norman)	25 01 04 01
	Arunella	subsalsa Brady	25 01 05 01
	Cyclocypris	globosa Sars	25 01 06 01
		laevis (Muller)	25 01 06 02
		ovum (Jurine)	25 01 06 03
		serena (Koch)	25 01 06 04
	Cypria	exsculpta (Fischer)	25 01 07 01
		ophthalmica (Jurine)	25 01 07 02

	Ilyocypris	*monstrifica* (Norman)	25 01 08 01
		gibba (Ramdohr)	25 01 08 02
		bradyi Sars	25 01 08 03
		decipiens Masi	25 01 08 04
		getica Masi	25 01 08 05
		inermis Kaufmann	25 01 08 06
	Notodromas	*monacha* (Muller)	25 01 09 01
	Cyprois	*marginata* (Strauss)	25 01 10 01
	Cypris	*bispinosa* Lucas	25 01 11 01
		pubera Muller	25 01 11 02
	Eucypris	*clavata* (Baird)	25 01 12 01
		crassa (Muller)	25 01 12 02
		elliptica (Baird)	25 01 12 03
		lutaria (Koch)	25 01 12 04
		ornata (Muller)	25 01 12 05
		pigra (Fischer)	25 01 12 06
		virens (Jurine)	25 01 12 07
		zenkeri (Chyzer)	25 01 12 08
		lilljeborgi Muller	25 01 12 09
		anglica Fose	25 01 12 10
	Cypricercus	*fuscatus* (Jurine)	25 01 13 01
		obliquus (Brady	25 01 13 02
		affinis (Fischer)	25 01 13 03
	Heterocypris	*incongruens* (Ramdohr)	25 01 14 01
		salina (Brady)	25 01 14 02
	Dolerocypris	*fasciata* (Muller)	25 01 15 01
	Herpetocypris	*brevicaudata* Kaufmann	25 01 16 01
		chevreuxi (Sars)	25 01 16 02
		intermedia Kaufmann	25 01 16 03
		palpiger Lowndes	25 01 16 04
		reptans (Baird)	25 01 16 05
		agilis Roma	25 01 16 06
	Ilyodromus	*olivaceus* (Brady & Norman)	25 01 17 01
		robertsoni (Brady & Norman)	25 01 17 02
	Scottia	*browniana* (Jones)	25 01 18 01
	Cypridopsis	*aculeata* (Costa)	25 01 19 01
		newtoni Brady & Robertson	25 01 19 02
		obesa (Brady & Robertson	25 01 19 03
		vidua (Muller)	25 01 19 04
		subterranea Wolf	25 01 19 05
	Potamocypris	*fulva* (Brady)	25 01 20 01
		pallida Alm	25 01 20 02
		variegata (Brady & Norman)	25 01 20 03
		villosa (Jurine)	25 01 20 04
		dianae Fose	25 01 20 05
		similis Muller	25 01 20 06
		fallase Fose	25 01 20 07
		wolfi Brehm	25 01 20 08
		thienemanii Klie	25 01 20 09
	Isocypris	*beauchampi* (Paris)	25 01 21 01
Darwinulidae	*Darwinula*	*stevensoni* (Brady & RObertson)	25 02 01 01
Cytheridae	*Limnocythere*	*inopinata* (Baird)	25 03 01 01
		sancti-patricii Brady & Robertson	25 03 01 02
	Cytherissa	*lacustris* Sars	25 03 02 01
	Leptocythere	*castanea* (Sars)	25 03 03 01
		pellucida (Baird)	25 03 03 02
	Metacypris	*cordata* Brady & Robertson	25 03 04 01
	Cyprideis	*torosa* (Jones)	25 03 05 01

		Cytheromorpha	*fuscata* (Brady)	25 03 06 01
26		**COPEPODA**		
	Centropagidae	*Centropages*	*hamatus* (Lilljeborg)	26 01 01 01
		Limnocalanus	*macrurus* Sars	26 01 02 01
	Diaptomidae	*Diaptomus*	*castor* (Jurine)	26 02 01 01
			laciniatus (Lilljeborg)	26 02 01 02
			gracilis (Sars)	26 02 01 03
			vulgaris (Schmeil)	26 02 01 04
			laticeps (Sars)	26 02 01 05
			wierzejskii (Richard)	26 02 01 06
	Temoridae	*Eurytemora*	*velox* Poppe	26 03 01 01
			affinis Lilljeborg	26 03 01 02
			americana Williams	26 03 01 03
	Acartiidae	*Acartia*	*clausi* Giesbrecht	26 04 01 01
			discaudata Giesbrecht	26 04 01 02
			bifilosa Giesbrecht	26 04 01 03
	Phyllognathopodidae	*Phyllognathopus*	*viguieri* (Maup)	26 05 01 01
	Tachidiidae	*Tachidius*	*discipes* Giesbrecht	26 06 01 01
			incisipes Klie	26 06 01 02
			littoralis Poppe	26 06 01 03
	Diosaccidae	*Stenhelia*	*palustris* (Brady)	26 07 01 01
	Canthocamptidae	*Nitocra*	*typica* Boek	26 08 01 01
			spinipes Boeck	26 08 01 02
			lacustris (Schmank)	26 08 01 03
			hibernica (Brady)	26 08 01 04
		Canthocamptus	*staphylinus* (Jurine)	26 08 02 01
			microstaphylinus Wolf	26 08 02 02
		Paracamptus	*schmeilii* (Mraz)	26 08 03 01
		Bryocamptus	*minutus* (Claus)	26 08 04 01
			pygmaeus (Sars)	26 08 04 02
			weberi (Kessl)	26 08 04 03
			typhlops (Kessl)	26 08 04 04
			zschokkei (Schmeil)	26 08 04 05
			cuspidatus (Schmeil)	26 08 04 06
			rhacticus (:Schmeil	26 08 04 07
		Echinocamptus	*echinatus* (Mraz)	26 08 05 01
			praegeri (Scourfield)	26 08 05 02
		Attheyella	*crassa* (Sars)	26 08 06 01
			bidens (Schmeil)	26 08 06 02
			dentata (Pogg)	26 08 06 03
			wulmeri (Kerh)	26 08 06 04
			trispinosus (Brady)	26 08 06 05
			gracilis (Sars)	26 08 06 06
		Moraria	*brevipes* (Sars)	26 08 07 01
			duthiei (Scott)	26 08 07 02
			poppei (Mrazek)	26 08 07 03
			mrazeki Scott	26 08 07 04
			varica (Graeter)	26 08 07 05
			arboricola Scourfield	26 08 07 06
			sphagnicola Gurney	26 08 07 07
		Mesochra	*lilljeborgii* Boek	26 08 08 01
			rapiens (Schmeil)	26 08 08 02
			aestuarii Gurney	26 08 08 03
		Maraenobiotus	*vejdovskyi* Mrazek	26 08 09 01
		Epactophanes	*richardi* Mrazek	26 08 10 01
			musicolus (Richters)	26 08 10 02
	Cylindropsyllidae	*Horsiella*	*brevicornis* Van Douwe	26 09 01 01
	Laophontidae	*Laophonte*	*mohammed* (Blanchard & Richard)	26 10 01 01

Cletodidae	Nannopus	*palustris* Brady	26 11 01 01
Cyclopinidae	Cyclopina	*norvegica* Boeck	26 12 01 01
Cyclopidae	Halicyclops	*aequoreus* Fischer	26 13 01 01
		neglectus Kiefer	26 13 01 02
	Macrocyclops	*fuscus* (Jurine)	26 13 02 01
		albidus (Jurine)	26 13 02 02
		distinctus (Richard)	26 13 02 03
	Tropocyclops	*prasinus* (Fischer)	26 13 03 01
	Eucyclops	*agilis* (Koch)	26 13 04 01
		speratus (Lilljeborg)	26 13 04 02
		macruroides (Lilljeborg)	26 13 04 03
		macrurus (Sars)	26 13 04 04
	Paracyclops	*fimbriatus* (Fischer)	26 13 05 01
		poppei (Rehberg)	26 13 05 02
		affinis (Sars)	26 13 05 03
	Ectocyclops	*phaleratus* (Koch)	26 13 06 01
	Cyclops	*strenuus* Fischer	26 13 07 01
		abyssorum Sars	26 13 07 02
		scutifer Sars	26 13 07 03
		furcifer Claus	26 13 07 04
		vicinus Uljanin	26 13 07 05
	Acanthocyclops	*viridis* (Jurine)	26 13 08 01
		gigas (Claus)	26 13 08 02
		latipes (Lowndes)	26 13 08 03
		vernalis (Fischer)	26 13 08 04
		americanus (Marsh)	26 13 08 05
		venustus (Norman & Scott)	26 13 08 06
		sensitivus (Graeter & Chappuis)	26 13 08 07
		bicuspidatus (Claus)	26 13 08 08
		bisetosus (Rehberg)	26 13 08 09
		crassicaudis (Sars)	26 13 08 10
		languidus (Sars)	26 13 08 11
		languidoides (Lilljeborg)	26 13 08 12
		nanus (Sars)	26 13 08 13
	Microcyclops	*varicans* (Sars)	26 13 09 01
		rubellus (Lilljeborg)	26 13 09 02
		bicolor (Sars)	26 13 09 03
		minutus (Claus)	26 13 09 04
		gracilis (Lilljeborg)	26 13 09 05
		unisetiger (Graeter)	26 13 09 06
		demetiensis (Scourfield)	26 13 09 07
	Mescyclops	*leuckarti* (Claus)	26 13 10 01
		hyalinus (Rehberg)	26 13 10 02
		dybowskii (Lands)	26 13 10 03
Ergasilidae	Ergasilus	*sieboldi* Nordmann	26 14 01 01
		gibbus Nordmann	26 14 01 02
	Thersitina	*gasterostei* Pagenstecher	26 14 02 01
Caligidae	Lepeophtheirus	*salmonis* (Kroyer)	26 15 01 01
Dichelesthiidae	Dichelasthium	*oblongum* Abildgaard	26 16 01 01
Lernaeidae	Lernaea	*cyprinacea* L.	26 17 01 01
Lernaeopodidae	Salmincola	*salmonea* (L.)	26 18 01 01
		thymalli (Kessler)	26 18 01 02
		gordoni Gurney	26 18 01 03
	Achtheres	*percarum* (Nordmann)	26 18 02 01
	Tracheliastes	*polycolpus* (Nordmann)	26 18 03 01

27   BRANCHIURA

| Argulidae | Argulus | *coregoni* Thorell | 27 01 01 01 |
| | | *foliaceus* (L.) | 27 01 01 02 |

28	MALACOSTRACA			
Bathynellidae	*Bathynella*	*natans* Vejdovsky	28 01 01 01
		stammeri Jakobi	28 01 01 02
Mysidae	*Mysis*	*relicta* Loven	28 02 01 01
	Neomysis	*integer* (Leach)	28 02 02 01
Asellidae	*Asellus*	*aquaticus* (L.)	28 03 01 01
		cavaticus Schiodte	28 03 01 02
		communis Say	28 03 01 03
		meridianus Racovitza	28 03 01 04
Sphaeromatidae	*Sphaeroma*	*hookeri* Leach	28 04 01 01
		rugicauda Leach	28 04 01 02
Janiridae	*Jaera*	*nordmanni* (Rathke)	28 05 01 01
Corophiidae	*Corophium*	*curvispinum* Sars	28 06 01 01
Gammaridae	*Crangonyx*	*pseudogracilis* Bousfield	28 07 01 01
		subterraneus Bate	28 07 01 02
	Echinogammarus	*berilloni* (Costa)	28 07 02 01
	Gammarus	*chevreuxi* Sexton	28 07 03 01
		duebeni Lilljeborg	28 07 03 02
		lacustris Sars	28 07 03 03
		locusta (L.)	28 07 03 04
		pulex (L.)	28 07 03 05
		oceanicus Segerstrale	28 07 03 06
		salinus Spooner	28 07 03 07
		tigrinus Sexton	28 07 03 08
		zaddachi Sexton	28 07 03 09
	Niphargellus	*glenniei* (Spooner)	28 07 04 01
	Niphargus	*aquilex* Schiodte	28 07 05 01
		fontanus Bate	28 07 05 02
		kochianus Bate	28 07 05 03
Talitridae	*Orchestia*	*cavimana* Heller	28 08 01 01
Palaemonidae	*Palaemonetes*	*varians* (Leach)	28 09 01 01
	Palaemon	*longirostris* Edwards	28 09 02 01
Astacidae	*Austropotamobius*	*pallipes* (Lereboullet)	28 10 01 01
Grapsidae	*Eriocheir*	*sinensis* Edwards	28 11 01 01
29 ARTHROPODA: INSECTA	COLLEMBOLA		
Poduridae	*Podura*	*aquatica* L.	29 01 01 01
Hypogastruridae	*Hypogastrura*	*viatica* (Tullberg)	29 02 01 01
	Anurida	*tullbergi* Schott	29 02 02 01
Isotomidae	*Proisotoma*	*crassicauda* (Tullberg)	29 03 01 01
	Ballistura	*schoetti* (Torre)	29 03 02 01
	Agrenia	*bidenticulata* (Tullberg)	29 03 03 01
	Isotoma	*viridis* Bourlet	29 03 04 01
		antennalis (Bagnall)	29 03 04 02
	Isotomurus	*alticolus* (Carl)	29 03 05 01
		palustris (Muller)	29 03 05 02
Smynthuridae	*Smithurides*	*aquaticus* (Bourlet)	29 04 01 01
		malmgreni (Tullberg)	29 04 01 02
		schoetti (Axelson)	29 04 01 03
		signatus (Krausbauer)	29 04 01 04
	Stenacidia	*violaceus* (Reuter)	29 04 02 01
	Heterosminthurus	*insignis* (Reuter)	29 04 03 01
		novemlineata (Tullberg)	29 04 03 02
30	EPHEMEROPTERA		
Siphlonuridae	*Siphlonurus*	*armatus* Eaton	30 01 01 01

		lacustris Eaton	30 01 01 02
		linnaeanus (Eaton)	30 01 01 03
	Ameletus	*inopinatus* Eaton	30 01 02 01
Baetidae	*Baetis*	*fuscatus* (L.)	30 02 01 01
		scambus Eaton	30 02 01 02
		vernus Curtis	30 02 01 03
		buceratus Eaton	30 02 01 04
		rhodani (Pictet)	30 02 01 05
		atrebatinus Eaton	30 02 01 06
		muticus (L.)	30 02 01 07
		niger (L.)	30 02 01 08
		digitatus Bengtsson	30 02 01 09
	Centroptilum	*luteolum* (Muller)	30 02 02 01
		pennulatum Eaton	30 02 02 02
	Cloeon	*dipterum* (L.)	30 02 03 01
		simile Eaton	30 02 03 02
	Procloeon	*pseudorufulum* Kimmins	30 02 04 01
Heptageniidae	*Rhithrogena*	*semicolorata* (Curtis)	30 03 01 01
		haarupi Esben-Petersen	30 03 01 02
	Heptagenia	*sulphurea* (Muller)	30 03 02 01
		longicauda (Stephens)	30 03 02 02
		fuscogrisea (Retzius)	30 03 02 03
		lateralis (Curtis)	30 03 02 04
	Arthroplea	*congener* Bengtsson	30 03 03 01
	Ecdyonurus	*venosus* (Fabricius)	30 03 04 01
		torrentis Kimmins	30 03 04 02
		dispar (Curtis)	30 03 04 03
		insignis (Eaton)	30 03 04 04
Leptophlebiidae	*Leptophlebia*	*marginata* (L.)	30 04 01 01
		vespertina (L.)	30 04 01 02
	Paraleptophlebia	*submarginata* (Stephens)	30 04 02 01
		cincta (Retzius)	30 04 02 02
		tumida Bengtsson	30 04 02 03
	Habrophlebia	*fusca* (Curtis)	30 04 03 01
Ephemerellidae	*Ephemerella*	*ignita* (Poda)	30 05 01 01
		notata Eaton	30 05 01 02
Potamanthidae	*Potamanthus*	*luteus* (L.)	30 06 01 01
Ephemeridae	*Ephemera*	*vulgata* L.	30 07 01 01
		danica Muller	30 07 01 02
		lineata Eaton	30 07 01 03
Caenidae	*Brachycercus*	*harrisella* Curtis	30 08 01 01
	Caenis	*macrura* Stephens	30 08 02 01
		moesta Bengtsson	30 08 02 02
		robusta Eaton	30 08 02 03
		horaria (L.)	30 08 02 04
		rivulorum Eaton	30 08 02 05

31	PLECOPTERA			
Taeniopterygidae	*Taeniopteryx*	*nebulosa* (L.)	31 01 01 01
	Rhabdiopteryx	*acuminata* Klapalek	31 01 02 01
	Brachyptera	*putata* (Newman)	31 01 03 01
		risi (Morton)	31 01 03 02
Nemouridae	*Protonemura*	*praecox* (Morton)	31 02 01 01
		montana Kimmins	31 02 01 02
		meyeri (Pictet)	31 02 01 03
	Amphinemura	*standfussi* Ris	31 02 02 01
		sulcicollis (Stephens)	31 02 02 02

	Nemurella	*picteti* Klapalek	31 02 03 01
	Nemoura	*cinerea* (Retzius)	31 02 04 01
		dubitans Morton	31 02 04 02
		avicularis Morton	31 02 04 03
		cambrica (Stephens)	31 02 04 04
		erratica Classen	31 02 04 05
Leuctridae	*Leuctra*	*geniculata* (Stephens)	31 03 01 01
		inermis Kempny	31 03 01 02
		hippopus (Kempny)	31 03 01 03
		nigra (Olivier)	31 03 01 04
		fusca (L.)	31 03 01 05
		moselyi Morton	31 03 01 06
Capniidae	*Capnia*	*bifrons* (Newman)	31 04 01 01
		atra Morton	31 04 01 02
		vidua Klapalek	31 04 01 03
Perlodidae	*Isogenus*	*nubecula* Newman	31 05 01 01
	Perlodes	*microcephala* (Pictet)	31 05 02 01
	Diura	*bicaudata* (L.)	31 05 03 01
	Isoperla	*grammatica* (Poda)	31 05 04 01
		obscura (Zetterstedt)	31 05 04 02
Perlidae	*Dinocras*	*cephalotes* (Curtis)	31 06 01 01
	Perla	*bipunctata* Pictet	31 06 02 01
Chloroperlidae	*Chloroperla*	*torrentium* (Pictet)	31 07 01 01
		tripunctata (Scopoli)	31 07 01 02
		apicalis Newman	31 07 01 03

32   ODONATA

Platycnemididae	*Platycnemis*	*pennipes* (Pallas)	32 01 01 01
Coenagriidae	*Pyrrhosoma*	*nymphula* (Sulzer)	32 02 01 01
	Ischnura	*elegans* (Linden)	32 02 02 01
		pumilio (Charpentier)	32 02 02 02
	Enallagma	*cyathigerum* (Charpentier)	32 02 03 01
	Coenagrion	*armatum* (Charpentier)	32 02 04 01
		hastulatum (Charpentier)	32 02 04 02
		mercuriale (Charpentier)	32 02 04 03
		puella (L.)	32 02 04 04
		pulchellum (Linden)	32 02 04 05
		scitulum (Rambur)	32 02 04 06
	Ceriagrion	*tenellum* (Villers)	32 02 05 01
	Erythromma	*najas* (Hansemann)	32 02 06 01
Lestidae	*Lestes*	*dryas* Kirby	32 03 01 01
		sponsa (Hansemann)	32 03 01 02
Agriidae	*Agrion*	*splendens* (Harris)	32 04 01 01
		virgo (L.)	32 04 01 02
Gomphidae	*Gomphus*	*vulgatissimus* (L.)	32 05 01 01
Cordulegasteridae	*Cordulegaster*	*boltonii* (Donovan)	32 06 01 01
Aeshnidae	*Brachytron*	*pratense* (Muller)	32 07 01 01
	Aeshna	*caerulea* (Strom)	32 07 02 01
		cyanea (Muller)	32 07 02 02
		grandis (L.)	32 07 02 03
		isosceles (Muller)	32 07 02 04
		juncea (L.)	32 07 02 05
		mixta Latreille	32 07 02 06
	Anax	*imperator* Leach	32 07 03 01
Corduliidae	*Cordulia*	*linaenea* Fraser	32 08 01 01
	Somatochlora	*alpestris* (Selys)	32 08 02 01
		artica (Zetterstedt)	32 08 02 02

		metallica (Linden)	32 08 02 03
	Oxygastra	curtisii (Dale)	32 08 03 01
Libellulidae	Orthetrum	cancellatum (L.)	32 09 01 01
		coerulescens (Fabricius)	32 09 01 02
	Libellula	depressa L.	32 09 02 01
		fulva Muller	32 09 02 02
		quadrimaculata L.	32 09 02 03
	Sympetrum	flaveolum (L.)	32 09 03 01
		fonscolombei (Selys)	32 09 03 02
		nigrescens Lucas	32 09 03 03
		sanguineum (Muller)	32 09 03 04
		scoticum (Donovan)	32 09 03 05
		striolatum (Charpentier)	32 09 03 06
		vulgatum (L.)	32 09 03 07
	Leucorrhinia	dubia (Linden)	32 09 04 01

33 HEMIPTERA

Mesovelidae	Mesovelia	furcata Mulsant & Rey	33 01 01 01
Hebridae	Hebrus	pusillus (Fallen)	33 02 01 01
		ruficeps (Thomson)	33 02 01 02
Hydrometridae	Hydrometra	gracilenta Horvath	33 03 01 01
		stagnorum (L.)	33 03 01 02
Veliidae	Velia	caprai Tamanini	33 04 01 01
		saulii Tamanini	33 04 01 02
	Microvelia	pygmaea (Dufour)	33 04 02 01
		reticulata (Burmeister)	33 04 02 02
		umbricola Wroblenski	33 04 02 03
Gerridae	Gerris	costai (Herrich-Schaffer)	33 05 01 01
		lateralis Schummel	33 05 01 02
		thoracicus Schummel	33 05 01 03
		gibbifer Schummel	33 05 01 04
		argentatus Schummel	33 05 01 05
		lacustris (L.)	33 05 01 06
		odontogaster (Zetterstedt)	33 05 01 07
		najas (Degeer)	33 05 01 08
		paludum (Fabricius)	33 05 01 09
		rufoscutellatus Latreille	33 05 01 10
Nepidae	Nepa	cinerea L.	33 06 01 01
	Ranatra	linearis (L.)	33 06 02 01
Naucoridae	Ilyocoris	cimicoides (L.)	33 07 01 01
Aphelocheiridae	Aphelocheirus	aestivalis (Fabricius)	33 08 01 01
Notonectidae	Notonecta	glauca L.	33 09 01 01
		viridis Delcourt	33 09 01 02
		obliqua Gallen	33 09 01 03
		maculata Fabricius	33 09 01 04
Pleidae	Plea	leachi McGregor & Kirkcaldy	33 10 01 01
Corixidae	Micronecta	scholtzi (Scholtz)	33 11 01 01
		minutissima (L.)	33 11 01 02
		poweri (Douglas & Scott)	33 11 01 03
	Cymatia	bonsdorffi (Sahlberg)	33 11 02 01
		coleoptrata (Fabricius)	33 11 02 02
	Glaenocorisa	propinqua (Fieber)	33 11 03 01
	Callicorixa	praeusta (Fieber)	33 11 04 01
		wollastoni (Douglas & Scott)	33 11 04 02
	Corixa	dentipes (Thomson)	33 11 05 01
		punctata (Illinger)	33 11 05 02
		affinis Leach	33 11 05 03

		panzeri (Fieber)	33 11 05 04
	Hesperocorixa	linnei (Fieber)	33 11 06 01
		sahlbergi (Fieber)	33 11 06 02
		castanea (Thomson)	33 11 06 03
		moesta (Fieber)	33 11 06 04
	Arctocorisa	carinata (Sahlberg)	33 11 07 01
		germari (Fieber)	33 11 07 02
	Sigara	dorsalis (Leach)	33 11 08 01
		striata (L.)	33 11 08 02
		distincta (Fieber)	33 11 08 03
		falleni (Fieber)	33 11 08 04
		fallenoidea (Hungerford)	33 11 08 05
		fossarum (Leach)	33 11 08 06
		scotti (Fieber)	33 11 08 07
		lateralis (Leach)	33 11 08 08
		nigrolineata (Fieber)	33 11 08 09
		concinna (Fieber)	33 11 08 10
		limitata (Fieber)	33 11 08 11
		semistriata (Fieber)	33 11 08 12
		venusta (Douglas & Scott)	33 11 08 13
		selecta (Fieber)	33 11 08 14
		stagnalis (Leach)	33 11 08 15

34 **HYMENOPTERA**

Braconidae	Ademon	decrescens (Nees)	34 01 01 01
	Opius	caesus Haliday	34 01 02 01
	Gyrocampa	affinis (Nees)	34 01 03 01
	Chorebus	uliginosus (Haliday)	34 01 04 01
	Chorebidea	naiadum (Haliday)	34 01 05 01
	Dacnusa	discolor Marshall	34 01 06 01
Agriotypidae	Agriotypus	armatus Curtis	34 02 01 01
Ichneumonidae	Rhembobius	perscrutator (Thunberg)	34 03 01 01
	Phygadeuon	vexator (Thunberg)	34 03 02 01
	Ischnurgops	lacustris (Schmiedeknecht)	34 03 03 01
	Hemiteles	cinctus (L.)	34 03 04 01
		persector Parfitt	34 03 04 02
		biannulatus Gravenhorst	34 03 04 03
Chalcididae	Chalcis	myrifex (Sulzer)	34 04 01 01
		sispes (L.)	34 04 01 02
Eurytomidae	Eudecatoma	biguttata (Swederus)	34 05 01 01
Torymidae	Monodontomerus	obscurus Westwood	34 06 01 01
Pteromalidae	Urolepis	maritimus (Walker)	34 07 01 01
	Cratomus	megacephalus (Fabricius)	34 07 02 01
	Cyrtogaster	clavicornis Walker	34 07 03 01
Trichogrammatidae	Trichogramma	evanescens Westwood	34 08 01 01
	Prestwichia	aquatica Lubbock	34 08 02 01
Mymaridae	Caraphractus	cinctus Walker	34 09 01 01
	Litus	cynipseus Haliday	34 09 02 01
Diapriidae	Diapria	conica (Fabricius)	34 10 01 01
	Paramesius	rufipes Westwood	34 10 02 01
Scelionidae	Thoron	metallicus (Curtis)	34 11 01 01

35 **COLEOPTERA**

| Haliplidae | Brychius | elevatus (Panzer) | 35 01 01 01 |
| | Peltodytes | caesus (Duftschmid) | 35 01 02 01 |

	Haliplus	*confinis* Stephens	35 01 03 01
		obliquus (Fabricius)	35 01 03 02
		lineatocollis (Marsham)	35 01 03 03
		ruficollis (Degeer)	35 01 03 04
		heydeni Wehncke	35 01 03 05
		fluviatilis Aube	35 01 03 06
		lineolatus Mannerheim	35 01 03 07
		immaculatus Gerhardt	35 01 03 08
		apicalis Thomson	35 01 03 09
		furcatus Seidlitz	35 01 03 10
		fulvus (Fabricius)	35 01 03 11
		flavicollis Sturm	35 01 03 12
		mucronatus Stephens	35 01 03 13
		variegatus Sturm	35 01 03 14
		laminatus Schaller	35 01 03 15
Hygrobiidae	*Hygrobia*	*hermanni* (Fabricius)	35 02 01 01
Dytiscidae	*Noterus*	*clavicornis* (Degeer)	35 03 01 01
	Laccophilus	*minutus* (L.)	35 03 02 01
		hyalinus (Degeer)	35 03 02 02
		variegatus (Germar)	35 03 02 03
	Hydrovatus	*clypealis* Sharp	35 03 03 01
	Hyphydrus	*ovatus* (L.)	35 03 04 01
	Bidessus	*unistriatus* (Schrank)	35 03 05 01
		minutissimus (Germar)	35 03 05 02
	Hygrotus	*inaequalis* (Fabricius)	35 04 06 01
		decoratus (Gyllenhal)	35 03 06 02
		versicolor (Schaller)	35 03 06 03
		quinquelineatus (Zetterstedt)	35 03 06 04
		confluens (Fabricius)	35 03 06 05
		novemlineatus (Stephens)	35 03 06 06
		parallelogrammus (Ahrens)	35 03 06 07
		impressopunctatus (Schaller)	35 03 06 08
	Deronectes	*latus* (Stephens)	35 03 07 01
		assimilis (Paykull)	35 03 07 02
		depressus (Fabricius)	35 03 07 03
		canariensis (Bedel)	35 03 07 04
		griseostriatus (Degeer)	35 03 07 05
	Oreodytes	*davisii* (Curtis)	35 03 08 01
		rivalis (Gyllenhal)	35 03 08 02
		septentrionalis (Gyllenhal)	35 03 08 03
	Hydroporus	*pictus* (Fabricius)	35 03 09 01
		granularis (L.)	35 03 09 02
		bilineatus Sturm	35 03 09 03
		flavipes (Olivier)	35 03 09 04
		lepidus (Olivier)	35 03 09 05
		dorsalis (Fabricius)	35 03 09 06
		lineatus (Fabricius)	35 03 09 07
		scalesianus Stephens	35 03 09 08
		neglectus Schaum	35 03 09 09
		umbrosus (Gyllenhal)	35 03 09 10
		angustatus Sturm	35 03 09 11
		morio Aube	35 03 09 12
		striola (Gyllenhal)	35 03 09 13
		palustris (L.)	35 03 09 14
		incognitus Sharp	35 03 09 15
		erythrocephalus (L.)	35 03 09 16
		rufifrons (Muller)	35 03 09 17
		longulus Mulsant	35 03 09 18
		longicornis Sharp	35 03 09 19
		melanarius Sturm	35 03 09 20
		memnonius Nicolai	35 03 09 21

		obscurus Sturm	35 03 09 22
		nigrita (Fabricius)	35 03 09 23
		discretus Fairmaire	35 03 09 24
		pubescens (Gyllenhal)	35 03 09 25
		planus (Fabricius)	35 03 09 26
		foveolatus (Heer)	35 03 09 27
		tesselatus Drapiez	35 03 09 28
		marginatus (Duftschmid)	35 03 09 29
		ferrugineus Stephens	35 03 09 30
		obsoletus Aube	35 03 09 31
	Laccornis	oblongus (Stephens)	35 03 10 01
	Agabus	guttatus (Paykull)	35 03 11 01
		biguttatus (Olivier)	35 03 11 02
		paludosus (Fabricius)	35 03 11 03
		brunneus (Fabricius)	35 03 11 04
		uliginosus (L.)	35 03 11 05
		affinis (Paykull)	35 03 11 06
		unguicularis Thomson	35 03 11 07
		didymus (Olivier)	35 03 11 08
		congener (Thunberg)	35 03 11 09
		nebulosus (Forster)	35 03 11 10
		conspersus (Marsham)	35 03 11 11
		striolatus (Gyllenhal)	35 03 11 12
		labiatus (Brahm)	35 03 11 13
		undulatus (Schrank)	35 03 11 14
		arcticus (Paykull)	35 03 11 15
		sturmii (Gyllenhal)	35 03 11 16
		chalconotus (Panzer)	35 03 11 17
		melanarius Aube	35 03 11 18
		bipustulatus (L.)	35 03 11 19
	Platambus	maculatus (L.)	35 03 12 01
	Ilybius	fuliginosus (Fabricius)	35 03 13 01
		subaeneus Erichson	35 03 13 02
		ater (Degeer)	35 03 13 03
		quadriguttatus (Lacordaire)	35 03 13 04
		guttiger (Gyllenhal)	35 03 13 05
		aenescens Thomson	35 03 13 06
		fenestratus (Fabricius)	35 03 13 07
	Copelatus	haemorrhoidalis (Fabricius)	35 03 14 01
	Rantus	grapii (Gyllenhal)	35 03 15 01
		exsoletus (Forster)	35 03 15 02
		pulverosus (Stephens)	35 03 15 03
		frontalis (Marsham)	35 03 15 04
		bistriatus (Bergstraesser)	35 03 15 05
	Colymbetes	fuscus (L.)	35 03 16 01
	Dytiscus	semisulcatus Muller	35 03 17 01
		marginalis L.	35 03 17 02
		circumflexus Fabricius	35 03 17 03
		circumcinctus Ahrens	35 03 17 04
		lapponicus Gyllenhal	35 03 17 05
		dimidiatus Bergstraesser	35 03 17 06
	Hydaticus	transversalis (Pontoppidan)	35 03 18 01
		seminiger (Degeer)	35 03 18 02
	Graphoderus	cinereus (L.)	35 03 19 01
	Acilius	sulcatus (L.)	35 03 20 01
Gyrinidae	Aulonogyrus	striatus (Fabricius)	35 04 01 01
	Gyrinus	minutus Fabricius	35 04 02 01
		urinotor Illiger	35 04 02 02
		natator (L.)	35 04 02 03
		suffriani Scriba	35 04 02 04
		bicolor Fabricius	35 04 02 05

		caspius Menetries	35 04 02 06
		colymbus Erichson	35 04 02 07
		marinus Gyllenhal	35 04 02 08
		aeratus Stephens	35 04 02 09
		opacus Sahlberg	35 04 02 10
	Orectochilus	*villosus* (Muller)	35 04 03 01
Hydrophilidae	*Ochthebius*	*exsculptus* Germar	35 05 01 01
		exaratus Mulsant	35 05 01 02
		dilatatus Stephens	35 05 01 03
		bicolon Germar	35 05 01 04
		auriculatus Rey	35 05 01 05
		minimus (Fabricius)	35 05 01 06
		aeneus Stephens	35 05 01 07
		punctatus Stephens	35 05 01 08
		nanus Stephens	35 05 01 09
		pusillus Stephens	35 05 01 10
		marinus (Paykull)	35 05 01 11
		lenensis Poppius	35 05 01 12
		viridis Peyron	35 05 01 13
	Hydraena	*testacea* Curtis	35 05 02 01
		palustris Erichson	35 05 02 02
		britteni Joy	35 05 02 03
		riparia Kugelmann	35 05 02 04
		nigrita Germar	35 05 02 05
		rufipes Curtis	35 05 02 06
		gracilis Germar	35 05 02 07
		pulchella Germar	35 05 02 08
		minutissima Stephens	35 05 02 09
		pygmaea Waterhouse	35 05 02 10
	Limnebius	*truncatellus* (Thunberg)	35 05 03 01
		nitidus (Marsham)	35 05 03 02
		aluta Bedel	35 05 03 03
	Spercheus	*emarginatus* (Schaller)	35 05 04 01
	Helophorus	*nubilus* Fabricius	35 05 05 01
		rufipes (Bosc d'Antic)	35 05 05 02
		porculus Bedel	35 05 05 03
		tuberculatus Gyllenhal	35 05 05 04
		alternans Gene	35 05 05 05
		aquaticus (L.)	35 05 05 06
		arvernicus Mulsant	35 05 05 07
		brevipalpis Bedel	35 05 05 08
		minutus Fabricius	35 05 05 09
		granularis (L.)	35 05 05 10
		flavipes Fabricius	35 05 05 11
		dorsalis (Marsham)	35 05 05 12
		laticollis Thomson	35 05 05 13
		nanus Sturm	35 05 05 14
	Hydrochus	*elongatus* (Schaller)	35 05 06 01
		carinatus Germar	35 05 06 02
		brevis (Herbst)	35 05 06 03
		nitidicollis Mulsant	35 05 06 04
		angustatus Germar	35 05 06 05
	Paracymus	*aeneus* (Germar)	35 05 07 01
		scutellaris (Rosenhauser)	35 05 07 02
	Hydrobius	*fuscipes* (L.)	35 05 08 01
	Limnoxenus	*niger* (Zschach)	35 05 09 01
	Anacaena	*globulus* (Paykull)	35 05 10 01
		limbata (Fabricius)	35 05 10 02
		bipustulata (Marsham)	35 05 10 03
	Laccobius	*minutus* (L.)	35 05 11 01
		biguttatus Gerhardt	35 05 11 02

		striatulus (Fabricius)	35 05 11 03
		sinuatus Motschulsky	35 05 11 04
		atratus Rottenburg	35 05 11 05
	Helochares	*lividus* (Forster)	35 05 12 01
	Enochrus	*melanocephalus* (Fabricius)	35 05 13 01
		bicolor (Fabricius)	35 05 13 02
		testaceus (Fabricius)	35 05 13 03
		affinis (Thunberg)	35 05 13 04
		ochropterus Marsham	35 05 13 05
	Cymbiodyta	*marginella* (Fabricius)	35 05 14 01
	Chaetarthia	*seminulum* (Herbst)	35 05 15 01
	Hydrochara	*caraboides* (L.)	35 05 16 01
	Hydrophilus	*piceus* (L.)	35 05 17 01
	Berosus	*spinosus* (von Steven)	35 05 18 01
		signaticollis Charpentier	35 05 18 02
		luridus (L.)	35 05 18 03
		affinis Brulle	35 05 18 04
Clambidae	*Calyptomerus*	*dubius* (Marsham)	35 06 01 01
	Clambus	*minutus* (Sturm)	35 06 02 01
		punctulum (Beck)	35 06 02 02
		armadillus (Degeer)	35 06 02 03
		pubescens Redtenbacher	35 06 02 04
Sphaeriidae	*Sphaerius*	*acaroides* Waltl	35 07 01 01
Dascillidae	*Eubria*	*palustris* Germar	35 08 01 01
Helodidae	*Helodes*	*minuta* (L.)	35 09 01 01
		marginata (Fabricius)	35 09 01 02
	Microcara	*testacea* (L.)	35 09 02 01
		bohemani (Mannerheim)	35 09 02 02
	Cyphon	*variabilis* (Thunberg)	35 09 03 01
		ochraceus Stephens	35 09 03 02
		padi (L.)	35 09 03 03
		coarctatus Paykull	35 09 03 04
		paykulli Guerin-Meneville	35 09 03 05
	Prionocyphon	*serricornis* (Muller)	35 09 04 01
	Hydrocyphon	*deflexicollis* (Muller)	35 09 05 01
	Scirtes	*hemisphaericus* (L.)	35 09 06 01
		orbicularis (Panzer)	35 09 06 02
Dryopidae	*Dryops*	*nitidulus* Heer	35 10 01 01
		ernesti DesGozis	35 10 01 02
		striatellus Fairmaire & Brisout	35 10 01 03
		luridus Erichson	35 10 01 04
		griseus Erichson	35 10 01 05
		anglicanus Edwards	35 10 01 06
		auriculatus (Geoffrey)	35 10 01 07
	Helichus	*substriatus* (Muller)	35 10 02 01
Elminthidae	*Elmis*	*aenea* (Muller)	35 11 01 01
	Esolus	*parallelepipedus* (Muller)	35 11 02 01
	Limnius	*volckmari* (Panzer)	35 11 03 01
	Macronychus	*quadrituberculatus* Muller	35 11 04 01
	Normandia	*nitens* (Muller)	35 11 05 01
	Oulimnius	*rivularis* (Rosenhauer)	35 11 06 01
		troglodytes (Gyllenhal)	35 11 06 02
		tuberculatus (Muller)	35 11 06 03
	Riolus	*cupreus* (Muller)	35 11 07 01
		subviolaceus (Muller)	35 11 07 02
	Stenelmis	*canaliculata* (Gyllenhal)	35 11 08 01

Georissidae	*Georissus*	*crenulatus* (Rossi)	35 12 01 01
Chrysomelidae	*Macroplea*	*appendiculata* (Panzer)	35 13 01 01
	Donacia	*clavipes* Fabricius	35 13 02 01
		crassipes Fabricius	35 13 02 02
		dentata Hoppe	35 13 02 03
		versicolorea (Brahm)	35 13 02 04
		semicuprea Panzer	35 13 02 05
		sparganii Ahrens	35 13 02 06
		aquatica (L.)	35 13 02 07
		impressa Paykull	35 13 02 08
		marginata Hoppe	35 13 02 09
		bicolor Zschach	35 13 02 10
		obscura Gyllenhal	35 13 02 11
		thalassima Germar	35 13 02 12
		vulgaris Zschach	35 13 02 13
		simplex Fabricius	35 13 02 14
		cinerea Herbst	35 13 02 15
	Plateumaris	*discolor* (Panzer)	35 13 03 01
		sericea (L.)	35 13 03 02
		braccata (Scopoli)	35 13 03 03
		affinis (Kunze)	35 13 03 04
Curculionidae	*Lixus*	*paraplecticus* (L.)	35 14 01 01
	Ephimeropus	*petro* (Herbst)	35 14 02 01
	Bagous	*cylindrus* (Paykull)	35 14 03 01
		binodulus (Herbst)	35 14 03 02
		nodulosus Gyllenhal	35 14 03 03
		argillaceus Gyllenhal	35 14 03 04
		lutulentus (Gyllenhal)	35 14 03 05
		glabrirostris (Herbst)	35 14 03 06
	Hydronomus	*alismatis* (Marsham)	35 14 04 01
	Tanysphyrus	*lemnae* (Paykull)	35 14 05 01
	Notaris	*bimaculatus* (Fabricius)	35 14 06 01
		scirpi (Fabricius)	35 14 06 02
		acridulus (L.)	35 14 06 03
	Thryogenes	*nereis* (Paykull)	35 14 07 01
		festucae (Herbst)	35 14 07 02
	Grypus	*equiseti* (Fabricius)	35 14 08 01
	Phytonomus	*adspersus* (Fabricius)	35 14 09 01
		rumicis (L.)	35 14 09 02
		arundinis (Paykull)	35 14 09 03
		suspiciosus (Herbst)	35 14 09 04
	Limnobaris	*t-album* (L.)	35 14 10 01
		pilistriata (Stephens)	35 14 10 02
	Mononychus	*punctum-album* (Herbst)	35 14 11 01
	Ceuthorhynchus	*melanostictus* (Marsham)	35 14 12 01
	Litodactylus	*leucogaster* (Marsham)	35 14 13 01
	Eubrychius	*velatus* (Beck)	35 14 14 01
	Drupenatus	*nasturtii* (Germar)	35 14 15 01
	Amalorrhynchus	*melanarius* (Stephens)	35 14 16 01
	Tapinotus	*sellatus* (Herbst)	35 14 17 01
	Poophagus	*sisymbrii* (Fabricius)	35 14 18 01
	Mecinus	*collaris* Germar	35 14 19 01
	Gymnetron	*villosulum* Gyllenhal	35 14 20 01
		beccabungae (L.)	35 14 20 02
		veronicae (Germar)	35 14 20 03
	Cionus	*alauda* (Herbst)	35 14 21 01

		scrophulariae (L.)	35 14 21 02
		hortulanus (Geoffroy)	35 14 21 03
	Cleopus	*pulchellus* (Herbst)	35 14 22 01
36	**MEGALOPTERA**		
Sialidae	*Sialis*	*lutaria* (L.)	36 01 01 01
		fuliginosa Pictet	36 01 01 02
37	**NEUROPTERA**		
Osmylidae	*Osmylus*	*fulvicephalus* (Scopoli)	37 01 01 01
Sisyridae	*Sisyra*	*fuscata* (Fabricius)	37 02 01 01
		dalii McLachlan	37 02 01 02
		terminalis Curtis	37 02 01 03
38	**TRICHOPTERA**		
Rhyacophilidae	*Rhyacophila*	*dorsalis* (Curtis)	38 01 01 01
		septentrionis McLachlan	38 01 01 02
		obliterata McLachlan	38 01 01 03
		munda McLachlan	38 01 01 04
	Glossosoma	*conformis* Neboiss	38 01 02 01
		boltoni Curtis	38 01 02 02
		intermedium (Klapalek)	38 01 02 03
	Agapetus	*fuscipes* Curtis	38 01 03 01
		ochripes Curtis	38 01 03 02
		delicatulus McLachlan	38 01 03 03
Philopotamidae	*Philopotamus*	*montanus* (Donovan)	38 02 01 01
	Wormaldia	*occipitalis* (Pictet)	38 02 02 01
		mediana McLachlan	38 02 02 02
		subnigra McLachlan	38 02 02 03
	Chimarra	*marginata* (L.)	38 02 03 01
Polycentropodidae	*Neureclipsis*	*bimaculata* (L.)	38 03 01 01
	Plectrocnemia	*conspersa* (Curtis)	38 03 02 01
		geniculata McLachlan	38 03 02 02
		brevis McLachlan	38 03 02 03
	Polycentropus	*flavomaculatus* (Pictet)	38 03 03 01
		irroratus (Curtis)	38 03 03 02
		kingi McLachlan	38 03 03 03
	Holocentropus	*dubius* (Rambur)	38 03 04 01
		picicornis (Stephens)	38 03 04 02
		stagnalis (Albarda)	38 03 04 03
	Cyrnus	*trimaculatus* (Curtis)	38 03 05 01
		insolutus McLachlan	38 03 05 02
		flavidus McLachlan	38 03 05 03
Psychomyiidae	*Ecnomus*	*tenellus* (Rambur)	38 04 01 01
	Tinodes	*waeneri* (L.)	38 04 02 01
		maclachlani Kimmins	38 04 02 02
		assimilis McLachlan	38 04 02 03
		pallidulus McLachlan	38 04 02 04
		maculicornis (Pictet)	38 04 02 05
		unicolor (Pictet)	38 04 02 06
		rostocki McLachlan	38 04 02 07
		dives (Pictet)	38 04 02 08
	Lype	*phaeopa* (Stephens)	38 04 03 01
		reducta (Hagen)	38 04 03 02
	Metalype	*fragilis* (Pictet)	38 04 04 01
	Psychomyia	*pusilla* (Fabricius)	38 04 05 01
Hydropsychidae	*Hydropsyche*	*pellucidula* (Curtis)	38 05 01 01
		angustipennis (Curtis)	38 05 01 02

		saxonica McLachlan	38 05 01 03
		contubernalis McLachlan	38 05 01 04
		guttata Pictet	38 05 01 05
		instabilis (Curtis)	38 05 01 06
		fulvipes (Curtis)	38 05 01 07
		exocellata Dufour	38 05 01 08
	Cheumatopsyche	*lepida* (Pictet)	38 05 02 01
	Diplectrona	*felix* McLachlan	38 05 03 01
Hydroptilidae	*Agraylea*	*mujltipunctata* Curtis	38 06 01 01
		sexmaculata Curtis	38 06 01 02
	Allotrichia	*pallicornis* (Eaton)	38 06 02 01
	Hydroptila	*sparsa* Curtis	38 06 03 01
		simulans Mosely	38 06 03 02
		cornuta Mosely	38 06 03 03
		lotensis Mosely	38 06 03 04
		angulata Mosely	38 06 03 05
		sylvestris Morton	38 06 03 06
		occulta (Eaton)	38 06 03 07
		tineoides Dalman	38 06 03 08
		pulchricornis Pictet	38 06 03 09
		forcipata (Eaton)	38 06 03 10
		vectis Curtis	38 06 03 11
		tigurina Ris	38 06 03 12
	Ithytrichia	*lamellaris* Eaton	38 06 04 01
		clavata Morton	38 06 04 02
	Orthotrichia	*angustella* (McLachlan)	38 06 05 01
		tragetti Mosely	38 06 05 02
		costalis (Curtis)	38 06 05 03
	Oxyethira	*flavicornis* (Pictet)	38 06 06 01
		tristella Klapalek	38 06 06 02
		simplex Ris	38 06 06 03
		falcata Morton	38 06 06 04
		frici Klapalek	38 06 06 05
		distinctella McLachlan	38 06 06 06
		sagittifera Ris	38 06 06 07
		mirabilis Morton	38 06 06 08
	Tricholeiochiton	*fagesi* (Guinard)	38 06 07 01
Phryganeidae	*Oligotricha*	*ruficrus* (Scopoli)	38 07 01 01
		clathrata (Kolenati)	38 07 01 02
	Phryganea	*grandis* L.	38 07 02 01
		striata L.	38 07 02 02
		varia Fabricius	38 07 02 03
		obsoleta McLachlan	38 07 02 04
	Trichostegia	*minor* (Curtis)	38 07 03 01
	Agrypnia	*picta* Kolenati	38 07 04 01
		pagetana Curtis	38 07 04 02
	Agrypnetes	*crassicornis* McLachlan	38 07 05 01
Limnephilidae	*Ironoquia*	*dubia* (Stephens)	38 08 01 01
	Apatania	*wallengreni* McLachlan	38 08 02 01
		auricula (Forsslund)	38 08 02 02
		muliebris McLachlan	38 08 02 03
	Drusus	*annulatus* Stephens	38 08 03 01
	Ecclisopteryx	*guttulata* (Pictet)	38 08 04 01
	Limnephilus	*rhombicus* (L.)	38 08 05 01
		flavicornis (Fabricius)	38 08 05 02
		subcentralis (Brauer)	38 08 05 03
		borealis (Zetterstedt)	38 08 05 04
		marmoratus Curtis	38 08 05 05
		politus McLachlan	38 08 05 06

		stigma Curtis	38 08 05 07
		binotatus Curtis	38 08 05 08
		decipiens Kolenati	38 08 05 09
		lanatus Curtis	38 08 05 10
		luridus Curtis	38 08 05 11
		ignavus McLachlan	38 08 05 12
		fuscinervis (Zetterstedt)	38 08 05 13
		elegans Curtis	38 08 05 14
		griseus (L.)	38 08 05 15
		bipunctatus Curtis	38 08 05 16
		affinis Curtis	38 08 05 17
		incisus Curtis	38 08 05 18
		hirsutus (Picet)	38 08 05 19
		centralis Curtis	38 08 05 20
		sparsus Curtis	38 08 05 21
		auricula Curtis	38 08 05 22
		vittatus (Fabricius)	38 08 05 23
		nigriceps Zetterstedt	38 08 05 24
		extricatus McLachlan	38 08 05 25
		fuscicornis (Rambur)	38 08 05 26
		coenosus Curtis	38 08 05 27
	Grammotaulius	*nitidus* (Muller)	38 08 06 01
		atomarius (Fabricius)	38 08 06 02
	Glyphotaelius	*pellucidus* (Retzius)	38 08 07 01
	Nemotaulius	*punctatolineatus* (Retzius)	38 08 08 01
	Anabolia	*nervosa* Curtis	38 08 09 01
		brevipennis (Curtis)	38 08 09 02
	Rhadicoleptus	*alpestris* (Kolenati)	38 08 10 01
	Potamophylax	*latipennis* (Curtis)	38 08 11 01
		cingulatus (Stephens)	38 08 11 02
		rotundipennis (Brauer)	38 08 11 03
	Halesus	*radiatus* (Curtis)	38 08 12 01
		digitatus (Schrank)	38 08 12 02
	Melampophylax	*mucoreus* (Hagen)	38 08 13 01
	Eniocyla	*pusilla* (Burmeister)	38 08 14 01
	Stenophylax	*permistus* McLachlan	38 08 15 01
		vibex (Curtis)	38 08 15 02
		lateralis (Stephens)	38 08 15 03
		sequax (McLachlan)	38 08 15 04
	Mesophylax	*impunctatus* McLachlan	38 08 16 01
		aspersus (Rambur)	38 08 16 02
	Allogamus	*auricollis* (Pictet)	38 08 17 01
	Hydatophylax	*infumatus* (McLachlan)	38 08 18 01
	Chaetopteryx	*villosa* (Fabricius)	38 08 19 01
Molannidae	*Molanna*	*angustata* Curtis	38 09 01 01
		palpata McLachlan	38 09 01 02
Beraeidae	*Beraea*	*pullata* (Curtis)	38 10 01 01
		maurus (Curtis)	38 10 01 02
	Ernodes	*articularis* (Pictet)	38 10 02 01
	Beraeodes	*minutus* (L.)	38 10 03 01
Odontoceridae	*Odontocerum*	*albicorne* (Scopoli)	38 11 01 01
Leptoceridae	*Athripsodes*	*nigronervosus* (Retzius)	38 12 01 01
		fulvus (Rambur)	38 12 01 02
		senilis (Burmeister)	38 12 01 03
		alboguttatus (Hagen)	38 12 01 04
		annulicornis (Stephens)	38 12 01 05
		aterrimus (Stephens)	38 12 01 06
		cinereus (Curtis)	38 12 01 07

		albifrons (L.)	38 12 01 08
		bilineatus (L.)	38 12 01 09
		commutatus (Rostock)	38 12 01 10
		dissimilis (Stephens)	38 12 01 11
	Mystacides	*nigra* (L.)	38 12 02 01
		azurea (L.)	38 12 02 02
		longicornis (L.)	38 12 02 03
	Triaenodes	*bicolor* (Curtis)	38 12 03 01
		conspersus (Rambur)	38 12 03 02
		simulans Tjeder	38 12 03 03
		reuteri McLachlan	38 12 03 04
	Erotesis	*baltica* McLachlan	38 12 04 01
	Adicella	*reducta* (McLachlan)	38 12 05 01
		filicornis (Pictet)	38 12 05 02
	Oecetis	*ochracea* (Curtis)	38 12 06 01
		furva (Rambur)	38 12 06 02
		lacustris (Pictet)	38 12 06 03
		notata (Rambur)	38 12 06 04
		testacea (Curtis)	38 12 06 05
	Leptocerus	*tineiformis* Curtis	38 12 07 01
		lusitanicus (McLachlan)	38 12 07 02
		interruptus (Fabricius)	38 12 07 03
	Setodes	*punctatus* (Fabricius)	38 12 08 01
		argentipunctellus McLachlan	38 12 08 02
Goeridae	*Goera*	*pilosa* (Fabricius)	38 13 01 01
	Silo	*pallipes* (Fabricius)	38 13 02 01
		nigricornis (Pictet)	38 13 02 02
Lepidostomatidae	*Crunoecia*	*irrorata* (Curtis)	38 14 01 01
	Lepidostoma	*hirtum* (Fabricius)	38 14 02 01
	Lasiocephala	*basalis* (Kolenati)	38 14 03 01
Brachycentridae	*Brachycentrus*	*subnubilus* Curtis	38 14 03 01
Sericostomatidae	*Sericostoma*	*personatum* (Spence)	38 15 01 01
	Notidobia	*ciliaris* (L.)	38 16 01 01

39 LEPIDOPTERA

Pyralidae	*Schoenobius*	*gigantella* (Denis&Schiffermuller)	39 01 01 01
		forficella (Thunberg)	39 01 01 02
	Donacaula	*mucronellus* (Denis&Schiffermuller)	39 01 02 01
	Acentria	*nivea* (Olivier)	39 01 03 01
	Nymphula	*nymphaeata* (L.)	39 01 04 01
	Parapoynx	*stratiotata* (L.)	39 01 05 01
		obscuralis (Grote)	39 01 05 02
		stagnata (Donovan)	39 01 05 03
	Cataclysta	*lemnata* (L.)	39 01 06 03
	Synclita	*obliteralis* (Walker)	39 01 07 01

40 DIPTERA

Tipulidae	*Prionocera*	*pubescens* Loew	40 01 01 01
		subserricornis (Zetterstedt)	40 01 01 02
		turcica (Fabricius)	40 01 01 03
	Dolichopeza	*albipes* (Strom)	40 01 02 01
	Ctenophora	*flaveolata* (Fabricius)	40 01 03 01
		ornata Meigen	40 01 03 02
		pectinicornis (L.)	40 01 03 03
	Nephrotoma	*aculeata* (Loew)	40 01 04 01
		analis Schummel)	40 01 04 02
		appendiculata (Pierre)	40 01 04 03

	cornicina (L.)	40	01	04	04
	crocata (L.)	40	01	04	05
	dorsalis (Fabricius)	40	01	04	06
	guestfalica (Westhoff)	40	01	04	07
	quadrifaria (Meigen)	40	01	04	08
Nigrotipula	*nigra* L.	40	01	05	01
Schummelia	*variicornis* Schummel	40	01	06	01
	yerburyi Edwards	40	01	06	02
Savtschenkia	*alpium* Bergroth	40	01	07	01
	cheethami Edwards	40	01	07	02
	gimmerthali Lackschevitz	40	01	07	03
	limbata Zetterstedt	40	01	07	04
	marmorata Meigen	40	01	07	05
	obsoleta Meigen	40	01	07	06
	pagana Meigen	40	01	07	07
	rufina Meigen	40	01	07	08
	serrulifera Alexander	40	01	07	09
	signata Staeger	40	01	07	10
	staegeri Nielsen	40	01	07	11
	subnodicornis Zetterstedt	40	01	07	12
Pterelachisus	*luridirostris* Schummel	40	01	08	01
	mutila Wahlgren	40	01	08	02
Beringotipula	*unca* Wiedemann	40	01	09	01
Dendrotipula	*flavolineata* Meigen	40	01	10	01
Mediotipula	*sarajevensis* Strobl	40	01	11	01
	siebkei Zetterstedt	40	01	11	02
Vestiplex	*montana* Curtis	40	01	12	01
Lindnerina	*bistilata* Lundstrom	40	01	13	01
Lunatipula	*helvola* Loew	40	01	14	01
Platytipula	*luteipennis* Meigen	40	01	15	01
	melanoceros Schummel	40	01	15	02
Yamatotipula	*coerulescens* Lackschevitz	40	01	16	01
	couckei Tonnoir	40	01	16	02
	lateralis Meigen	40	01	16	03
	marginata Meigen	40	01	16	04
	montium Egger	40	01	16	05
	pruinosa Wiedemann	40	01	16	06
	solstitialis Westhoff	40	01	16	07
Tipula	*czizeki* de Jong	40	01	17	01
	oleracea L.	40	01	17	02
	paludosa Meigen	40	01	17	03
Acutipula	*fulvipennis* Degeer	40	01	18	01
	luna Westhoff	40	01	18	02
	maxima Poda	40	01	18	03
	vittata Meigen	40	01	18	04
Triogma	*trisulcata* (Schummel)	40	01	19	01
Phalacrocera	*replicata* (L.)	40	01	20	01
Limonia	*flavipes* (Fabricius)	40	01	21	01
	nubeculosa Meigen	40	01	21	02
Dicranomyia	*autumnalis* (Staeger)	40	01	22	01
	chorea (Meigen)	40	01	22	02
	didyma Meigen	40	01	22	03
	mitis (Meigen)	40	01	22	04
	modesta (Meigen)	40	01	22	05
Geranomyia	*bezzii* (Alexander & Leonard)	40	01	23	01
	unicolor (Haliday)	40	01	23	02
Rhipidia	*duplicata* (Doane)	40	01	24	01
Antocha	*vitripennis* Meigen	40	01	25	01

Thaumastoptera	*calceata* Mik	40 01 26 01
Orimarga	*juvenilis* (Zetterstedt)	40 01 27 01
Elliptera	*omissa* Schiner	40 01 28 01
Helius	*longirostris* (Meigen)	40 01 29 01
Pedicia	*rivosa* (L.)	40 01 30 01
Crunobia	*straminea* (Meigen)	40 01 31 01
Amalopis	*occulata* (Meigen)	40 01 32 01
Tricyphona	*immaculata* (Meigen)	40 01 33 01
	schummeli (Edwards)	40 01 33 02
Ludicia	*claripennis* (Verrall)	40 01 34 01
Dicranota	*bimaculata* (Schummel)	40 01 35 01
	guerini Zetterstedt	40 01 35 02
Paradicranota	*exclusa* (Walker)	40 01 36 01
	subtilis Loew	40 01 36 02
Oxyrhiza	*senilis* (Haliday)	40 01 37 01
Austrolimnophila	*ochracea* (Meigen)	40 01 38 01
Dactylolabis	*sexmaculata* (Macquart)	40 01 39 01
	transversa (Meigen)	40 01 39 02
Pseudolimnophila	*lucorum* (Meigen)	40 01 40 01
Eloeophila	*maculata* (Meigen)	40 01 41 01
	trimaculata (Zetterstedt)	40 01 41 02
	verralli (Bergroth)	40 01 41 03
Phylidorea	*ferruginea* (Meigen)	40 01 42 01
	lineola (Meigen)	40 01 42 02
	meigeni Verrall	40 01 42 03
Limnophila	*pictipennis* (Meigen)	40 01 43 01
	punctata (Schrank)	40 01 43 02
Brachylimnophila	*nemoralis* (Meigen)	40 01 44 01
Pilaria	*discicollis* (Meigen)	40 01 45 01
	fuscipennis (Meigen)	40 01 45 02
Ellipteroides	*lateralis* (Macquart)	40 01 46 01
Gonomyia	*dentata* de Meijere	40 01 47 01
	tenella (Meigen)	40 01 47 02
Lipsothrix	*remota* (Walker)	40 01 48 01
Platytoma	*cinerascens* (Meigen)	40 01 49 01
Trimicra	*pilipes* (Fabricius)	40 01 50 01
Helobia	*hybrida* (Meigen)	40 01 51 01
	stictica (Meigen)	40 01 51 02
Erioptera	*divisa* (Walker)	40 01 52 01
	fuscipennis Meigen	40 01 52 02
	gemina Tjeder	40 01 52 03
	lutea Meigen	40 01 52 04
	trivialis Meigen	40 01 52 05
Ilisia	*areolata* Siebke	40 01 53 01
	maculata Miegen	40 01 53 02
Ormosia	*hederae* (Curtis)	40 01 54 01
	lineata (Meigen)	40 01 54 02
Rhypholophus	*haemorrhoidalis* (Zetterstedt)	40 01 55 01
	varia (Meigen)	40 01 55 02
Psychodidae *Sycorax*	*silacea* Curtis	40 02 01 01
Pericoma	*blandula* Eaton	40 02 02 01
	calcilega Feuerborn	40 02 02 02
	canescens (Meigen)	40 02 02 03
	cognata Eaton	40 02 02 04

		compta Eaton	40 02 02 05
		crispi Freeman	40 02 02 06
		diversa Tonnoir	40 02 02 07
		exquisita Eaton	40 02 02 08
		extricata Eaton	40 02 02 09
		fallax Eaton	40 02 02 10
		fuliginosa (Meigen)	40 02 02 11
		gracilis Eaton	40 02 02 12
		hibernica Tonnoir	40 02 02 13
		mutua Eaton	40 02 02 14
		neglecta Eaton	40 02 02 15
		neoblandula Duckhouse	40 02 02 16
		nubila (Meigen)	40 02 02 17
		palustris (Meigen)	40 02 02 18
		pilularia Tonnoir	40 02 02 19
		pseudoexquisita Tonnoir	40 02 02 20
		pulchra Eaton	40 02 02 21
		trifasciata (Meigen)	40 02 02 22
		trivialis Eaton	40 02 02 23
	Telmatoscopus	advenus (Eaton)	40 02 03 01
		ambiguus (Eaton)	40 02 03 02
		britteni Tonnoir	40 02 03 03
		consors (Eaton)	40 02 03 04
		decipiens (Eaton)	40 02 03 05
		fratercula (Eaton)	40 02 03 06
		labeculosus (Eaton)	40 02 03 07
		laurencei Freeman	40 02 03 08
		longicornis (Tonnoir)	40 02 03 09
		mooni Duckhouse	40 02 03 10
		morulus (Eaton)	40 02 03 11
		soleatus (Walker)	40 02 03 12
		sylviae Duckhouse	40 02 03 13
		angustipennis (Tonnoir)	40 02 03 14
		rothschildi Eaton	40 02 03 15
		tristis (Meigen)	40 02 03 16
		ustulata (Walker)	40 02 03 17
	Peripsychoda	auriculata (Curtis)	40 02 04 01
		fusca (Macquart)	40 02 04 02
	Mormia	andrenipes (Strobl)	40 02 05 01
		caliginosa (Eaton)	40 02 05 02
		eatoni (Tonnoir)	40 02 05 03
		furva (Tonnoir)	40 02 05 04
		incerta (Eaton)	40 02 05 05
		palposa (Tonnoir)	40 02 05 06
		revisenda (Eaton)	40 02 05 07
	Panimerus	albifacies (Tonnoir)	40 02 06 01
		goetghebueri (Tonnoir)	40 02 06 02
		notabilis (Eaton)	40 02 06 03
	Threticus	lucifugus (Walker)	40 02 07 01
	Clytocerus	dalii (Eaton)	40 02 08 01
		ocellaris (Meigen)	40 02 08 02
	Psychoda	alternata Say	40 02 09 01
		cinerea Banks	40 02 09 02
		gemina Eaton	40 02 09 03
		obscura Tonnoir	40 02 09 04
		phalaenoides (L.)	40 02 09 05
		severini Tonnoir	40 02 09 06
Ptychopteridae	Ptychoptera	albimana (Fabricius)	40 03 01 01
		contaminata (L.)	40 03 01 02
		lacustris Meigen	40 03 01 03
		longicauda (Tonnoir)	40 03 01 04
		minuta Tonnoir	40 03 01 05
		paludosa Meigen	40 03 01 06

		scutellaris Meigen	40	03	01	07
Dixidae	*Dixa*	*dilatata* Strobl	40	04	01	01
		maculata Meigen	40	04	01	02
		nebulosa Meigen	40	04	01	03
		nubilipennis Curtis	40	04	01	04
		puberula Loew	40	04	01	05
		submaculata Edwards	40	04	01	06
	Dixella	*aestivalis* Meigen	40	04	02	01
		amphibia Degeer	40	04	02	02
		attica Pandazis	40	04	02	03
		autumnalis Meigen	40	04	02	04
		filicornis Edwards	40	04	02	05
		martinii Peus	40	04	02	06
		obscura Loew	40	04	02	07
		serotina Meigen	40	04	02	08
Chaoboridae	*Chaoborus*	*crystallinus* (Degeer)	40	05	01	01
		flavicans (Meigen)	40	05	01	02
		obscuripes (Wulp)	40	05	01	03
	Peusomyia	*pallidus* (Fabricius)	40	05	02	01
	Mochlonyx	*culiciformis* (Degeer)	40	05	03	01
		fuliginosus (Felt)	40	05	03	02
Culicidae	*Anopheles*	*algeriensis* Theobald	40	06	01	01
		atroparvus van Thiel	40	06	01	02
		claviger (Meigen)	40	06	01	03
		messeae Falleroni	40	06	01	04
		plumbeus Stephens	40	06	01	05
	Mansonia	*richiardii* (Ficalbi)	40	06	02	01
	Orthopodomyia	*pulchripalpis* (Rondani)	40	06	03	01
	Ochlerotatus	*annulipes* (Meigen)	40	06	04	01
		cantans (Meigen)	40	06	04	02
		caspius (Pallas)	40	06	04	03
		communis (Degeer)	40	06	04	04
		detritus (Haliday)	40	06	04	05
		dorsalis (Meigen)	40	06	04	06
		flavescens (Muller)	40	06	04	07
		leucomelas (Meigen)	40	06	04	08
		punctor (Kirby)	40	06	04	09
		rusticus (Rossi)	40	06	04	10
		sticticus (Meigen)	40	06	04	11
	Finlaya	*geniculatus* (Olivier)	40	06	05	01
	Stegomyia	*aegypti* (L.)	40	06	06	01
	Aedimorphus	*vexans* (Meigen)	40	06	07	01
	Aedes	*cinereus* Meigen	40	06	08	01
	Culiseta	*alaskaensis* (Ludlow)	40	06	09	01
		annulata (Schrank)	40	06	09	02
		subochrea (Edwards)	40	06	09	03
	Culicella	*fumipennis* (Stephens)	40	06	10	01
		litorea (Shute)	40	06	10	02
		morsitans (Theobald)	40	06	10	03
	Allotheobaldia	*longiareolata* (Macquart)	40	06	11	01
	Barraudius	*modestus* Ficalbi	40	06	12	01
	Neoculex	*territans* Walker	40	06	13	01
	Culex	*pipiens* L.	40	06	14	01
		torrentium Martini	40	06	14	02
Thaumaleidae	*Thaumalea*	*testacea* Ruthe	40	07	01	01
		truncata Edwards	40	07	01	02
		verralli Edwards	40	07	01	03
Ceratopogonidae	*Euforcipomyia*	*crassipes* (Winnertz)	40	08	01	01

	titillans (Winnertz)	40 08 01 02
Caloforcipomyia	*glauca* Macfie	40 08 02 01
Forcipomyia	*bipunctata* (Linnaeus)	40 08 03 01
	braueri (Wasmann)	40 08 03 02
	brevipennis (Macquart)	40 08 03 03
	ciliata (Winnertz)	40 08 03 04
	kaltenbachii (Winnertz)	40 08 03 05
	myrmecophila (Egger)	40 08 03 06
	nigra (Winnertz)	40 08 03 07
	picea (Winnertz)	40 08 03 08
	pulchrithorax Edwards	40 08 03 09
	radicicola Edwards	40 08 03 10
	regulus (Winnertz)	40 08 03 11
	solonensis Wirth	40 08 03 12
	tenuisquama Kieffer	40 08 03 13
Microhelea	*fuliginosa* (Meigen)	40 08 04 01
Pterobosca	*paludis* Macfie	40 08 05 01
Synthyridomyia	*murina* (Winnertz)	40 08 06 01
Thyridomyia	*frutetorum* (Winnertz)	40 08 07 01
	monilicornis Coquillett	40 08 07 02
	rugosa Chan & Le Roux	40 08 07 03
Trichohelea	*eques* (Johannsen)	40 08 08 01
	papilionivora Edwards	40 08 08 02
Atrichopogon	*aethiops* (Goetghebuer)	40 08 09 01
	appendiculatus (Goetghebuer)	40 08 09 02
	brunnipes (Meigen)	40 08 09 03
	fuscus (Meigen)	40 08 09 04
	hamifer (Goetghebuer)	40 08 09 05
	lucorum (Meigen)	40 08 09 06
	minutus (Meigen)	40 08 09 07
	pavidus (Winnertz)	40 08 09 08
	pollinivorus Downes	40 08 09 09
	rostratus (Winnertz)	40 08 09 10
	trifasciatus Kieffer	40 08 09 11
	winnertzi Goetghebuer	40 08 09 12
Meloehelea	*meloesugans* Kieffer	40 08 10 01
	oedemerarum Stora	40 08 10 02
Dasyhelea	*bensoni* Edwards	40 08 11 01
	dufouri (Laboulbene)	40 08 11 02
	flavifrons (Guerin-Meneville)	40 08 11 03
	flaviventris Goetghebuer	40 08 11 04
	flavoscutellata (Zetterstedt)	40 08 11 05
	holosericea (Meigen)	40 08 11 06
	lithotelmatica Strenzke	40 08 11 07
	notata Goetghebuer	40 08 11 08
	polita Edwards	40 08 11 09
	saxicola (Edwards)	40 08 11 10
	scutellata (Meigen)	40 08 11 11
	versicolor (Winnertz)	40 08 11 12
Avaritia	*chiopterus* (Meigen)	40 08 12 01
	dewulfi Goetghebuer	40 08 12 02
	obsoletus (Meigen)	40 08 12 03
Beltranmyia	*circumscriptus* Kieffer	40 08 13 01
	manchuriensis Tokunaga	40 08 13 02
	salinarius Kieffer	40 08 13 03
Culicoides	*delta* Edwards	40 08 14 01
	fagineus Edwards	40 08 14 02
	grisescens Edwards	40 08 14 03
	halophilus Kieffer	40 08 14 04
	impunctatus Goetghebuer	40 08 14 05

	lupicaris Downes & Kettle	40 08 14 06
	pulicaris (Linnaeus)	40 08 14 07
	punctatus (Meigen)	40 08 14 08
Monoculicoides	*nubeculosus* (Meigen)	40 08 15 01
	parroti Kieffer	40 08 15 02
	puncticollis (Becker)	40 08 15 03
	riethi Kieffer	40 08 15 04
	stigma (Meigen)	40 08 15 05
Oecacta	*achrayi* Kettle & Lawson	40 08 16 01
	albicans (Winnertz)	40 08 16 02
	brunnicans Edwards	40 08 16 03
	cataneii Clastrier	40 08 16 04
	duddingstoni Kettle & Lawson	40 08 16 05
	fascipennis (Staeger)	40 08 16 06
	furcillatus Callot,Kremer&Paradis	40 08 16 07
	heliophilus Edwards	40 08 16 08
	kibunensis Tokunaga	40 08 16 09
	maritimus Kieffer	40 08 16 10
	minutissimus (Zetterstedt)	40 08 16 11
	musilator Kremer & Callot	40 08 16 12
	odibilis Austen	40 08 16 13
	pallidicornis Kieffer	40 08 16 14
	pictipennis (Staeger)	40 08 16 15
	picturatus Kremer & Deduit	40 08 16 16
	poperinghensis Goetghebuer	40 08 16 17
	saevanicus Dzhafarov	40 08 16 18
	simulator Edwards	40 08 16 19
	subfasciipennis Kieffer	40 08 16 20
	truncorum Edwards	40 08 16 21
	vexans (Staeger)	40 08 16 22
Wirthomyia	*reconditus* Campbell & Pelham-Clinton	40 08 17 01
	segnis Campbell & Pelham-Clinton	40 08 17 02
	cameroni Campbell & Pelham-Clinton	40 08 17 03
Ceratopogon	*communis* Meigen	40 08 18 01
	crassinervis Goetghebuer	40 08 18 02
	lacteipennis Zetterstedt	40 08 18 03
	niveipennis Meigen	40 08 18 04
Isohelea	*nitidula* (Edwards)	40 08 19 01
	perpusilla (Edwards)	40 08 19 02
	sociabilis (Goetghebuer)	40 08 19 03
Monohelea	*tessellata* (Zetterstedt)	40 08 20 01
Schizohelea	*leucopeza* (Meigen)	40 08 21 01
Serromyia	*femorata* (Meigen)	40 08 22 01
	morio (Fabricius)	40 08 22 02
	nitens Goetghebuer	40 08 22 03
Neostilobezzia	*calcarata* (Goetghebuer)	40 08 23 01
	gracilis (Haliday)	40 08 23 02
	lutacea Edwards	40 08 23 03
	ochracea (Winnertz)	40 08 23 04
	sharpi Edwards	40 08 23 05
Stilobezzia	*flavirostris* (Winnertz)	40 08 24 01
Clinohelea	*unimaculata* (Macquart)	40 08 25 01
Neurohelea	*luteitarsis* (Meigen)	40 08 26 01
Mallochohelea	*inermis* (Kieffer)	40 08 27 01
	munda (Loew)	40 08 27 02
	nitida (Macquart)	40 08 27 03
	setigera (Loew)	40 08 27 04
Probezzia	*seminigra* (Panzer)	40 08 28 01

	Sphaeromias	*candidatus* (Loew)	40 08 29 01
		fasciatus (Meigen)	40 08 29 02
	Palpomyia	*armipes* (Meigen)	40 08 30 01
		brevicornis Edwards	40 08 30 02
		distincta (Haliday)	40 08 30 03
		ephippium (Zetterstedt)	40 08 30 04
		flavipes (Meigen)	40 08 30 05
		fulva (Macquart)	40 08 30 06
		grossipes Goetghebuer	40 08 30 07
		lineata (Meigen)	40 08 30 08
		luteifemorata Edwards	40 08 30 09
		nemorivaga Goetghebuer	40 08 30 10
		nigripes (Meigen)	40 08 30 11
		praeusta (Loew)	40 08 30 12
		quadrispinosa Goetghebuer	40 08 30 13
		semifumosa (Goetghebuer)	40 08 30 14
		serripes (Meigen)	40 08 30 15
		spinipes (Meigen)	40 08 30 16
	Bezzia	*albipes* (Winnertz)	40 08 31 01
		annulipes (Meigen)	40 08 31 02
		bicolor (Meigen)	40 08 31 03
		calceata (Walker)	40 08 31 04
		decincta Edwards	40 08 31 05
		flavicornis (Staeger)	40 08 31 06
		gracilis (Winnertz)	40 08 31 07
		multiannulata (Strobl)	40 08 31 08
		nigritula (Zetterstedt)	40 08 31 09
		nobilis (Winnertz)	40 08 31 10
		ornata (Meigen)	40 08 31 11
		pygmaea Goetghebuer	40 08 31 12
		rubiginosa (Winnertz)	40 08 31 13
		solstitialis (Winnertz)	40 08 31 14
		taeniata (Walker)	40 08 31 15
		xanthocephala Goetghebuer	40 08 31 16
Chironomidae (Tanypodinae)	*Tanypus*	*punctipennis* Meigen	40 09 01 01
		vilipennis (Kieffer)	40 09 01 02
	Apsectrotanypus	*trifascipennis* (Zetterstedt)	40 09 02 01
	Macropelopia	*goetghebueri* (Keiffer)	40 09 03 01
		nebulosa (Meigen)	40 09 03 02
		notata (Meigen)	40 09 03 03
	Procladius	*choreus* (Meigen)	40 09 04 01
		crassinervis (Zetterstedt)	40 09 04 02
		culciformis (L.)	40 09 04 03
		sagittalis (Kieffer)	40 09 04 04
		simplicistilus Freeman	40 09 04 05
	Psectrotanypus	*varius* (Fabricius)	40 09 05 01
	Psilotanypus	*flavifrons* (Edwards)	40 09 06 01
		lugens (Kieffer)	40 09 06 02
		rufovittatus (Wulp)	40 09 06 03
	Clinotanypus	*nervosus* (Meigen)	40 09 07 01
	Ablabesmyia	*longistyla* Fittkau	40 09 08 01
		monilis (Linnaeus)	40 09 08 02
		phatta (Egger)	40 09 08 03
	Arctopelopia	*barbitarsis* (Zetterstedt)	40 09 09 01
		griseipennis (Wulp)	40 09 09 02
	Conchapelopia	*melanops* (Meigen)	40 09 10 01
		pallidula (Meigen)	40 09 10 02
		viator (Kieffer)	40 09 10 03
	Guttipelopia	*guttipennis* (Wulp)	40 09 11 01
	Krenopelopia	*binotata* (Wiedemann)	40 09 12 01

		nigropunctata (Staeger)	40 09 12 02
		schineri (Strobl)	40 09 12 03
	Ladbrundinia	longipalpis (Goetghebuer)	40 09 13 01
	Monopelopia	tenuicalcar (Keiffer)	40 09 14 01
	Natarsia	nugax (Walker)	40 09 15 01
		punctata (Fabricius)	40 09 15 02
	Nilotanypus	dubius (Meigen)	40 09 16 01
	Paramerina	cingulata (Walker)	40 09 17 01
		divisa (Walker)	40 09 17 02
	Rheopelopia	eximia (Edwards)	40 09 18 01
		maculipennis (Zetterstedt)	40 09 18 02
		ornata (Meigen)	40 09 18 03
Telmatopelopia		nemorum (Goetghebuer)	40 09 19 01
Thienemannimyia		carnea (Fabricius)	40 09 20 01
		fusciceps (Edwards)	40 09 20 02
		laeta (Meigen)	40 09 20 03
		lentiginosa (Fries)	40 09 20 04
		northumbrica (Edwards)	40 09 20 05
		woodi (Edwards)	40 09 20 06
	Trissopelopia	longimana (Staeger)	40 09 21 01
	Xenopelopia	falcigera (Kieffer)	40 09 22 01
		nigricans (Goetghebuer)	40 09 22 02
	Zavrelimyia	barbatipes (Kieffer)	40 09 23 01
		hirtimana (Kieffer)	40 09 23 02
		melanura (Meigen)	40 09 23 03
		nubila (Meigen)	40 09 23 04
(Podonominae)	Lasiodiamesa	sphagnicola (Kieffer)	40 10 01 01
	Parochlus	kiefferi (Garrett)	40 10 02 01
(Diamesinae)	Protanypus	morio (Zetterstedt)	40 11 01 01
	Diamesa	bohemani Goetghebuer	40 11 02 01
		incallida (Walker)	40 11 02 02
		insignipes Kieffer	40 11 02 03
		latitarsis (Goetghebuer)	40 11 02 04
		permacra (Walker)	40 11 02 05
		thienemanni Kieffer	40 11 02 06
		tonsa (Walker)	40 11 02 07
	Potthastia	gaedii (Meigen)	40 11 03 01
		longimana Kieffer	40 11 03 02
		montium (Edwards)	40 11 03 03
		pastoris (Edwards)	40 11 03 04
	Pseudodiamesa	branickii (Nowicki)	40 11 04 01
		nivosa (Goetghebuer)	40 11 04 02
	Pseudokiefferiella	parva (Edwards)	40 11 05 01
	Sympotthastia	zavreli Pagast	40 11 06 01
	Syndiamesa	edwardsi (Pagast)	40 11 07 01
	Thalassomya	frauenfeldi Schiner	40 11 08 01
(Telmatogetoninae)	Psamathiomya	pectinata (Deby)	40 12 01 01
(Orthocladiinae)	Acricotopus	lucens (Zetterstedt)	40 13 01 01
	Brillia	longifurca Kieffer	40 13 02 01
		modesta (Meigen)	40 13 02 02
	Cardiocladius	capucinus (Zetterstedt)	40 13 03 01
		fuscus (Kieffer)	40 13 03 02
	Cricotopus	albiforceps (Kieffer)	40 13 04 01
		annulator Goetghebuer	40 13 04 02
		bicinctus (Meigen)	40 13 04 03
		ephippium (Zetterstedt)	40 13 04 04

	festivellus (Kieffer)	40 13 04 05
	flavocinctus (Kieffer)	40 13 04 06
	fuscus (Kieffer)	40 13 04 07
	lygropis Edwards	40 13 04 08
	pallidipes Edwards	40 13 04 09
	pilosellus Brundin	40 13 04 10
	polaris (Kieffer)	40 13 04 11
	pulchripes Verrall	40 13 04 12
	similis Goetghebuer	40 13 04 13
	tremulus (L.)	40 13 04 14
	triannulatus (Macquart)	40 13 04 15
	trifascia Edwards	40 13 04 16
Isocladius	*brevipalpis* Kieffer	40 13 05 01
	intersectus (Staeger)	40 13 05 02
	laricomalis (Edwards)	40 13 05 03
	obnixus (Walker)	40 13 05 04
	ornatus (Meigen)	40 13 05 05
	pilitarsis (Zetterstedt)	40 13 05 06
	reversus Hirvenoja	40 13 05 07
	speciosus Goetghebuer	40 13 05 08
	sylvestris (Fabricius)	40 13 05 09
	tricinctus (Meigen)	40 13 05 10
	trifasciatus (Panzer)	40 13 05 11
Diplocladius	*cultriger* Kieffer	40 13 06 01
Eukiefferiella	*brevicalcar* (Kieffer)	40 13 07 01
	calvescens (Edwards)	40 13 07 02
	claripennis (Lundbeck)	40 13 07 03
	clypeata (Kieffer)	40 13 07 04
	coerulescens (Kieffer)	40 13 07 05
	devonica (Edwards)	40 13 07 06
	discoloripes Goetghebuer	40 13 07 07
	gracei (Edwards)	40 13 07 08
	ilkleyensis (Edwards)	40 13 07 09
	minor (Edwards)	40 13 07 10
	verralli (Edwards)	40 13 07 11
Eurycnemus	*crassipes* (Panzer)	40 13 08 01
Halocladius	*fucicola* (Edwards)	40 13 09 01
	variabilis (Staeger)	40 13 09 02
	varians (Staeger)	40 13 09 03
Heterotanytarsus	*apicalis* (Kieffer)	40 13 10 01
Heterotrissocladius	*grimshawi* (Edwards)	40 13 11 01
	marcidus (Walker)	40 13 11 02
Microcricotopus	*bicolor* (Zetterstedt)	40 13 12 01
	rectinervis (Kieffer)	40 13 12 02
Monodiamesa	*bathyphila* (Kieffer)	40 13 13 01
Odontomesa	*fulva* (Kieffer)	40 13 14 01
Eudactylocladius	*gelidus* Kieffer	40 13 15 01
	obtexens Brundin	40 13 15 02
	scanicus Brundin	40 13 15 03
Pogonocladius	*consobrinus* (Holmgren)	40 13 16 01
Euorthocladius	*frigidus* (Zetterstedt)	40 13 17 01
	rivicola Kieffer	40 13 17 02
	thienemanni Kieffer	40 13 17 03
Orthocladius	*dentifer* Brundin	40 13 18 01
	distalis Goetghebuer	40 13 18 02
	flaveolus Goetghebuer	40 13 18 03
	oblidens (Walker)	40 13 18 04
	rhyacobius Kieffer	40 13 18 05
	rubicundus (Meigen)	40 13 18 06
	saxicola Kieffer	40 13 18 07

Paracladius	*conversus* (Walker)	40 13 19 01
Paratrichocladius	*rufiventris* (Meigen)	40 13 20 01
	skirwithensis (Edwards)	40 13 20 02
Prodiamesa	*olivacea* (Meigen)	40 13 21 01
	rufovittata Goetghebuer	40 13 21 02
Monopsectrocladius	*calcaratus* (Edwards)	40 13 22 01
Allopsectrocladius	*obvius* (Walker)	40 13 23 01
	platypus (Edwards)	40 13 23 02
Psectrocladius	*barbimanus* (Edwards)	40 13 24 01
	edwardsi Brundin	40 13 24 02
	fennicus Stora	40 13 24 03
	limbatellus (Holmgren)	40 13 24 04
	psilopterus Kieffer	40 13 24 05
	sordidellus (Zetterstedt)	40 13 24 06
	turfaceus Kieffer	40 13 24 07
	ventricosus Kieffer	40 13 24 08
Rheocricotopus	*atripes* (Kieffer)	40 13 25 01
	chalybeatus (Edwards)	40 13 25 02
	dispar (Goetghebuer)	40 13 25 03
	effusus (Walker)	40 13 25 04
	glabricollis (Meigen)	40 13 25 05
Synorthocladius	*semivirens* (Kieffer)	40 13 26 01
Acamptocladius	*submontanus* (Edwards)	40 13 27 01
Bryophaenocladius	*aestivus* Brundin	40 13 28 01
	femineus (Edwards)	40 13 28 02
	furcatus (Kieffer)	40 13 28 03
	ictericus (Meigen)	40 13 28 04
	illimbatus (Edwards)	40 13 28 05
	nidorum (Edwards)	40 13 28 06
	nitidicollis (Goetghebuer)	40 13 28 07
	simus (Edwards)	40 13 28 08
	subvernalis (Edwards)	40 13 28 09
	tuberculatus (Edwards)	40 13 28 10
	vernalis (Goetghebuer)	40 13 28 11
	xanthogyne (Edwards)	40 13 28 12
Camptocladius	*stercorarius* (Degeer)	40 13 29 01
Chaetocladius	*dentiforceps* (Edwards)	40 13 30 01
	dissipatus (Edwards)	40 13 30 02
	melaleucus (Meigen)	40 13 30 03
	perennis (Meigen)	40 13 30 04
	piger (Goetghebuer)	40 13 30 05
	suecicus (Kieffer)	40 13 30 06
Clunio	*marinus* Haliday	40 13 31 01
Corynoneura	*carriana* Edwards	40 13 32 01
	celeripes Winnertz	40 13 32 02
	celtica Edwards	40 13 32 03
	coronata Edwards	40 13 32 04
	edwardsi Brundin	40 13 32 05
	fuscihalter Edwards	40 13 32 06
	lacustris Edwards	40 13 32 07
	lobata Edwards	40 13 32 08
	minutissimus (Meigen)	40 13 32 09
	scutellata Winnertz	40 13 32 10
Epoicocladius	*flavens* (Malloch)	40 13 33 01
Gymnometriocnemus	*brevitarsis* (Edwards)	40 13 34 01
	brumalis (Edwards)	40 13 34 02
	subnudus (Edwards)	40 13 34 03
Heleniella	*ornaticollis* (Edwards)	40 13 35 01
Krenosmittia	*camptophleps* (Edwards)	40 13 36 01

	Limnophyes	exiguus (Goetghebuer)	40	13	37	01
		gurgicola (Edwards)	40	13	37	02
		habilis (Walker)	40	13	37	03
		minimus (Meigen)	40	13	37	04
		prolongatus (Kieffer)	40	13	37	05
		pumilio (Holmgren)	40	13	37	06
		truncorum (Goetghebuer)	40	13	37	07
	Mesosmittia	flexuella (Edwards)	40	13	38	01
	Metriocnemus	atratulus (Zetterstedt)	40	13	39	01
		atriclavus Kieffer	40	13	39	02
		fuscipes (Meigen)	40	13	39	03
		gracei Edwards	40	13	39	04
		hirticollis (Staeger)	40	13	39	05
		hygropetricus (Kieffer)	40	13	39	06
		martinii Thienemann	40	13	39	07
		picipes (Meigen)	40	13	39	08
		tristellus Edwards	40	13	39	09
		ursinis (Holmgren)	40	13	39	10
	Orthosmittia	albipennis (Goetghebuer)	40	13	40	01
		brevifurcata (Edwards)	40	13	40	02
	Parakiefferiella	bathophila (Kieffer)	40	13	41	01
		coronata (Edwards)	40	13	41	02
	Paralimnophyes	hydrophilus (Goetghebuer)	40	13	42	01
	Parametriocnemus	stylatus (Kieffer)	40	13	43	01
	Paraphaenocladius	cuneatus (Edwards)	40	13	44	01
		impensus (Walker)	40	13	44	02
		irritus (Walker)	40	13	44	03
		penerasus (Edwards)	40	13	44	04
	Paratrissocladius	exerptus (Walker)	40	13	45	01
	Pseudorthocladius	curtistylus (Goetghebuer)	40	13	46	01
		filiformis (Kieffer)	40	13	46	02
		pilosipennis Brundin	40	13	46	03
	Pseudosmittia	angusta (Edwards)	40	13	47	01
		conjuncta (Edwards)	40	13	47	02
		curticosta (Edwards)	40	13	47	03
		forcipata (Goetghebuer)	40	13	47	04
		gracilis (Goetghebuer)	40	13	47	05
		recta (Edwards)	40	13	47	06
		scotica (Edwards)	40	13	47	07
		trilobata (Edwards)	40	13	47	08
	Smittia	aterrima (Meigen)	40	13	48	01
		contigens (Walker)	40	13	48	02
		edwardsi Goetghebuer	40	13	48	03
		foliacea (Kieffer)	40	13	48	04
		leucopogon (Meigen)	40	13	48	05
		nudopennis (Goetghebuer)	40	13	48	06
		pratorum (Goetghebuer)	40	13	48	07
		superata Goetghebuer	40	13	48	08
	Thalassosmittia	thalassophila (Goetghebuer)	40	13	49	01
	Thienemannia	gracilis Kieffer	40	13	50	01
	Thienemanniella	clavicornis (Kieffer)	40	13	51	01
		flavescens (Edwards)	40	13	51	02
		lutea (Edwards)	40	13	51	03
		majuscula (Edwards)	40	13	51	04
		morosa (Edwards)	40	13	51	05
		vittata (Edwards)	40	13	51	06
(Chironominae)	Camptochironomus	pallidivittatus (Malloch)	40	14	01	01
		tentans (Fabricius)	40	14	01	02
	Chironomus	annularis (Degeer)	40	14	02	01

	anthracinus Zetterstedt	40 14 02 02
	aprilinus Meigen	40 14 02 03
	cingulatus Meigen	40 14 02 04
	dorsalis Meigen	40 14 02 05
	halophilus Kieffer	40 14 02 06
	inermifrons Goetghebuer	40 14 02 07
	longistylus Goetghebuer	40 14 02 08
	lugubris Zetterstedt	40 14 02 09
	luridus Strenzke	40 14 02 10
	obtusidens Goetghebuer	40 14 02 11
	pilicornis (Fabricius)	40 14 02 12
	plumosus (L.)	40 14 02 13
	pseudothummi Strenzke	40 14 02 14
	riparius Meigen	40 14 02 15
	salinarius Kieffer	40 14 02 16
	striatus Strenzke	40 14 02 17
Cryptochironomus	*albofasciatus* (Staeger)	40 14 03 01
	denticulatus Goetghebuer	40 14 03 02
	obreptans (Walker)	40 14 03 03
	psittacinus (Meigen)	40 14 03 04
	redekei (Krusemann)	40 14 03 05
	rostratus Kieffer	40 14 03 06
	supplicans (Meigen)	40 14 03 07
Cryptocladopelma	*edwardsi* (Kruseman)	40 14 04 01
	krusemani (Goetghebuer)	40 14 04 02
	laccophila (Kieffer)	40 14 04 03
	lateralis (Goetghebuer)	40 14 04 04
	virescens (Meigen)	40 14 04 05
	viridula (L.)	40 14 04 06
Cryptotendipes	*nigronitens* (Edwards)	40 14 05 01
	psuedotener (Goetghebuer)	40 14 05 02
Demeijerea	*rufipes* (L.)	40 14 06 01
Demicryptochironomus	*vulneratus* (Zetterstedt)	40 14 07 01
Einfeldia	*dissidens* (Walker)	40 14 08 01
	longipes (Staeger)	40 14 08 02
	macani (Freeman)	40 14 08 03
	paganus (Meigen)	40 14 08 04
Endochironomus	*albipennis* (Meigen)	40 14 09 01
	dispar (Meigen)	40 14 09 02
	impar (Walker)	40 14 09 03
	intextus (Walker)	40 14 09 04
	lepidus (Meigen)	40 14 09 05
	tendens (Fabricius)	40 14 09 06
Glyptotendipes	*barbipes* (Staeger)	40 14 10 01
	foliicola Kieffer	40 14 10 02
	gripekoveni (Kieffer)	40 14 10 03
	mancunianus (Edwards)	40 14 10 04
	pallens (Meigen)	40 14 10 05
	paripes (Edwards)	40 14 10 06
	severini (Goetghebuer)	40 14 10 07
	viridis (Macquart)	40 14 10 08
Graceus	*ambiguus* Goetghebuer	40 14 11 01
Harnischia	*curtilamèllatus* (Malloch)	40 14 12 01
Kiefferulus	*tendipediformis* (Goetghebuer)	40 14 13 01
Lauterborniella	*agrayloides* (Kieffer)	40 14 14 01
Leptochironomus	*tener* (Kieffer)	40 14 15 01
Limnochironomus	*lobiger* Kieffer	40 14 16 01
	nervosus (Staeger)	40 14 16 02
	notatus (Meigen)	40 14 16 03
	pulsus (Walker)	40 14 16 04

	tritomus (Kieffer)	40	14	16	05
Microtendipes	*britteni* (Edwards)	40	14	17	01
	caledonicus (Edwards)	40	14	17	02
	chloris (Meigen)	40	14	17	03
	confinis (Meigen)	40	14	17	04
	diffinis (Edwards)	40	14	17	05
	lugubris Kieffer	40	14	17	06
	nitidus (Meigen)	40	14	17	07
	pedellus (Degeer)	40	14	17	08
	rydalensis (Edwards)	40	14	17	09
	tarsalis (Walker)	40	14	17	10
Nilothauma	*brayi* (Goetghebuer)	40	14	18	01
Pagastiella	*orophilus* (Edwards)	40	14	19	01
Parachironomus	*arcuatus* (Goetghebuer)	40	14	20	01
	biannulatus (Staeger)	40	14	20	02
	digitalis (Edwards)	40	14	20	03
	frequens (Johanssen)	40	14	20	04
	mauricii (Kruseman)	40	14	20	05
	monochromus (Wulp)	40	14	20	06
	parilis (Walker)	40	14	20	07
	tenuicaudatus (Malloch)	40	14	20	08
	varus (Goetghebuer)	40	14	20	09
	vitiosus (Geotghebuer)	40	14	20	10
Paracladopelma	*camptolabis* (Kieffer)	40	14	21	01
	obscura Brundin	40	14	21	02
Paralauterborniella	*nigrohalteralis* (Malloch)	40	14	22	01
Paratendipes	*albimanus* (Meigen)	40	14	23	01
	nudisquama (Edwards)	40	14	23	02
Pentapedilum	*nubens* (Edwards)	40	14	24	01
	sordens (Wulp)	40	14	24	02
	tritum (Walker)	40	14	24	03
	uncinatum Goetghebuer	40	14	24	04
Phaenopsectra	*flavipes* (Meigen)	40	14	25	01
	punctipes (Wiedemann)	40	14	25	02
Polypedilum	*acutum* Kieffer	40	14	26	01
	albicornis (Meigen)	40	14	26	02
	apfelbecki (Strobl)	40	14	26	03
	arundineti (Goetghebuer)	40	14	26	04
	bicrenatum Kieffer	40	14	26	05
	convictum (Walker)	40	14	26	06
	cultellatum Goetghebuer	40	14	26	07
	laetum (Meigen)	40	14	26	08
	leucopus (Meigen)	40	14	26	09
	nubeculosum (Meigen)	40	14	26	10
	pedestris (Meigen)	40	14	26	11
	pullum (Zetterstedt)	40	14	26	12
	quadriguttatum Kieffer	40	14	26	13
	scalaenum (Schrank)	40	14	26	14
Pseudochironomus	*prasinatus* (Staeger)	40	14	27	01
Sergentia	*coracina* (Zetterstedt)	40	14	28	01
Stenochironomus	*fascipennis* (Zetterstedt)	40	14	29	01
	gibbus (Fabricius)	40	14	29	02
	hibernicus (Edwards)	40	14	29	03
Stictochironomus	*maculipennis* (Meigen)	40	14	30	01
	pictulus (Meigen)	40	14	30	02
	rosenschoeldi (Zetterstedt)	40	14	30	03
	sticticus (Fabricius)	40	14	30	04
Xenochironomus	*xenolabis* (Kieffer)	40	14	31	01
Zavreliella	*marmorata* (Wulp)	40	14	32	01
Cladotanytarsus	*atridorsum* Kieffer	40	14	33	01

	difficilis Brundin	40 14 33 02
	mancus (Walker)	40 14 33 03
	nigrovittatus (Goetghebuer)	40 14 33 04
	vandewulpi (Edwards)	40 14 33 05
Micropsectra	*apposita* (Walker)	40 14 34 01
	atrofasciata Kieffer	40 14 34 02
	attenuata Reis	40 14 34 03
	bidentata (Goetghebuer)	40 14 34 04
	dentatiloba Kieffer	40 14 34 05
	foliata Laville	40 14 34 06
	fusca (Meigen)	40 14 34 07
	groenlandica Andersen	40 14 34 08
	praecox (Meigen)	40 14 34 09
	recurvata (Goetghebuer)	40 14 34 10
	solitaria Goetghebuer	40 14 34 11
	subnitens Goetghebuer	40 14 34 12
	subviridis (Goetghebuer)	40 14 34 13
	tenellula (Goetghebuer)	40 14 34 14
Parapsectra	*chionophila* (Edwards)	40 14 35 01
	nana (Meigen)	40 14 35 02
Paratanytarsus	*austriacus* Kieffer	40 14 36 01
	bituberculatus (Edwards)	40 14 36 02
	inopertus (Walker)	40 14 36 03
	intricatus (Goetghebuer)	40 14 36 04
	laccophilus (Edwards)	40 14 36 05
	laetipes (Zetterstedt)	40 14 36 06
	natvigi (Goetghebuer)	40 14 36 07
	penicillatus (Goetghebuer)	40 14 36 08
	tenuis (Meigen)	40 14 36 09
Tanytarsus	*arduennensis* Goetghebuer	40 14 37 01
	bathophilus Kieffer	40 14 37 02
	brundini Lindeberg	40 14 37 03
	buchonius Reiss & Fittkau	40 14 37 04
	debilis (Meigen)	40 14 37 05
	ejuncidus (Walker)	40 14 37 06
	eminulus (Walker)	40 14 37 07
	excavatus Edwards	40 14 37 08
	fimbriatus Reiss & Fittkau	40 14 37 09
	giabrescens Edwards	40 14 37 10
	gracilentus Holmgren	40 14 37 11
	gregarius Kieffer	40 14 37 12
	heusdensis Goetghebuer	40 14 37 13
	holochlorus Edwards	40 14 37 14
	inaequalis Goetghebuer	40 14 37 15
	lactescens Edwards	40 14 37 16
	lestagei Goetghebuer	40 14 37 17
	lugens Kieffer	40 14 37 18
	miriforceps (Kieffer)	40 14 37 19
	nemorosus Edwards	40 14 37 20
	norvegicus Kieffer	40 14 37 21
	palettaris Verneaux	40 14 37 22
	pallidicornis (Walker)	40 14 37 23
	quadridentatus Brundin	40 14 37 24
	signatus (Wulp)	40 14 37 25
	sylvaticus (Wulp)	40 14 37 26
	triangularis Goetghebuer	40 14 37 27
	usmaensis Pagast	40 14 37 28
	verralli Goetghebuer	40 14 37 29
Rheotanytarsus	*curtistylus* (Goetghebuer)	40 14 38 01
	muscicola Kieffer	40 14 38 02
	pentapoda Kieffer	40 14 38 03
	photophilus (Goetghebuer)	40 14 38 04
	reissi Lehmann	40 14 38 05

		ringei Lehmann	40 14 38 06
	Stempellina	*bausei* (Kieffer)	40 14 39 01
	Stempellinella	*brevis* (Edwards)	40 14 40 01
		cuneipennis (Edwards)	40 14 40 02
		flavidula (Edwards)	40 14 40 03
		minor (Edwards)	40 14 40 04
	Zavrelia	*pentatoma* Kieffer	40 14 41 01
Simuliidae	*Prosimulium*	*hirtipes* Fries	40 15 01 01
		arvernense Grenier	40 15 01 02
		inflatum Davies	40 15 01 03
	Cnephia	*tredecimatum* Edwards	40 15 02 01
		amphora Ladle & Bass	40 15 02 02
	Eusimulium	*subexcisum* Edwards	40 15 03 01
		yerburyi Edwards	40 15 03 02
		latipes Meigen	40 15 03 03
		brevicaule Dorier & Grenier	40 15 03 04
		armoricanum Doby & David	40 15 03 05
		dunfellense Davies	40 15 03 06
		naturale Davies	40 15 03 07
		urbanum Davies	40 15 03 08
		costatum Friederichs	40 15 03 09
		angustitarse Lundstrom	40 15 03 10
		latigonium Rubtzov	40 15 03 11
		cambriense Davies	40 15 03 12
		aureum Fries	40 15 03 13
		angustipes Edwards	40 15 03 14
	Wilhelmia	*equinum* L.	40 15 04 01
		zetlandense Davies	40 15 04 02
		salopiense Edwards	40 15 04 03
	Simulium	*tuberosum* Lundstrom	40 15 05 01
		reptans L.	40 15 05 02
		morsitans Edwards	40 15 05 03
		austeni Edwards	40 15 05 04
		sublacustre Davies	40 15 05 05
		argyreatum Meigen	40 15 05 06
		erythrocephalum Degeer	40 15 05 07
		monticola Friederichs	40 15 05 08
		variegatum Meigen	40 15 05 09
		ornatum Meigen	40 15 05 10
		nitidifrons Edwards	40 15 05 11
		spinosum Doby & Deblock	40 15 05 12
		britannicum Davies	40 15 05 13
Stratiomyidae	*Beris*	*clavipes* (L.)	40 16 01 01
		vallata (Forster)	40 16 01 02
	Nemotelus	*nigrinus* Fallen	40 16 02 01
		notatus Zetterstedt	40 16 02 02
		pantherinus (L.)	40 16 02 03
		uliginosus (L.)	40 16 02 04
	Oxycera	*analis* Meigen	40 16 03 01
		dives Loew	40 16 03 02
		formosa Meigen	40 16 03 03
		morrisii Curtis	40 16 03 04
		nigripes Verrall	40 16 03 05
		pardalina Meigen	40 16 03 06
		pulchella Meigen	40 16 03 07
		pygmaea (Fallen)	40 16 03 08
		terminata Meigen	40 16 03 09
		trilineata (Fabricius)	40 16 03 10
	Vanoyia	*tenuicornis* Macquart	40 16 04 01
	Odontomyia	*angulata* (Panzer)	40 16 05 01

		argentata (Fabricius)	40 16 05 02
		hydroleon (L.)	40 16 05 03
		ornata (Meigen)	40 16 05 04
		tigrina (Fabricius)	40 16 05 05
		viridula (Fabricius)	40 16 05 06
	Stratiomys	*chamaeleon* (L.)	40 16 06 01
		furcata (Fabricius)	40 16 06 02
		longicornis (Scopoli)	40 16 06 03
		potamida (Meigen)	40 16 06 04
Empididae	*Phyllodromia*	*melanocephala* (Fabricius)	40 17 01 01
	Chelipoda	*albiseta* (Zetterstedt)	40 17 02 01
		vocatoria (Fallen)	40 17 02 02
	Chelifera	*angusta* Collin	40 17 03 01
		aperticauda Collin	40 17 03 02
		astigma Collin	40 17 03 03
		concinnicauda Collin	40 17 03 04
		diversicauda Collin	40 17 03 05
		flavella (Zetterstedt)	40 17 03 06
		monostigma (Meigen)	40 17 03 07
		pectinicauda Collin	40 17 03 08
		precabunda Collin	40 17 03 09
		precatoria (Fallen)	40 17 03 10
		stigmatica Schiner	40 17 03 11
		subangusta Collin	40 17 03 12
		trapezina (Zetterstedt)	40 17 03 13
	Hemerodromia	*adulatoria* Collin	40 17 04 01
		baetica Collin	40 17 04 02
		laudatoria Collin	40 17 04 03
		melangyna Collin	40 17 04 04
		oratoria (Fallen)	40 17 04 05
		raptoria Meigen	40 17 04 06
		unilineata Zetterstedt	40 17 04 07
	Dryodromia	*testacea* Rondani	40 17 05 01
	Heleodromia	*immaculata* Haliday	40 17 06 01
	Trichopeza	*longicornis* (Meigen)	40 17 07 01
	Dolichocephala	*guttata* (Haliday)	40 17 08 01
		irrorata (Fallen)	40 17 08 02
		ocellata (Costa)	40 17 08 03
	Atalanta	*bipunctata* (Haliday)	40 17 09 01
		fontinalis Haliday	40 17 09 02
		nigra (Loew)	40 17 09 03
		nivalis (Zetterstedt)	40 17 09 04
		stagnalis (Haliday)	40 17 09 05
		tenella (Wahlberg)	40 17 09 06
		wesmaeli (Macquart)	40 17 09 07
	Wiedemannia	*bistigma* (Curtis)	40 17 10 01
		fallaciosa (Mik)	40 17 10 02
		lamellata (Loew)	40 17 10 03
		lota Haliday	40 17 10 04
		phantasma (Mik)	40 17 10 05
		rhynchops Collin	40 17 10 06
	Stilpon	*graminum* (Fallen)	40 17 11 01
		lunata (Walker)	40 17 11 02
		nubila Collin	40 17 11 03
		sublunata Collin	40 17 11 04
	Rhamphomyia	*barbata* (Macquart)	40 17 12 01
		flava (Fallen)	40 17 12 02
Dolichopodidae	*Dolichopus*	*atratus* (Meigen)	40 18 01 01
		cilifemoratus (Macquart)	40 18 01 02
		griseipennis (Stannius)	40 18 01 03

		nubilus (Meigen)	40 18 01 04
		pennatus (Meigen)	40 18 01 05
		plumipes (Scopoli)	40 18 01 07
		ungulatus (L.)	40 18 01 08
	Hercostomus	*aerosus* (Fallen)	40 18 02 01
		assimilis (Staeger)	40 18 02 02
		chetifer (Walker)	40 18 02 03
	Tachytrechus	*insignis* Stannius	40 18 03 01
		notatus Stannius	40 18 03 02
	Hydrophorus	*balticus* (Meigen)	40 18 04 01
		bipunctatus (Lehmann)	40 18 04 02
		borealis (Loew)	40 18 04 03
		litoreus (Fallen)	40 18 04 04
		nebulosus (Zetterstedt)	40 18 04 05
		oceanus (Macquart)	40 18 04 06
		praecox (Lehmann)	40 18 04 07
		viridis (Meigen)	40 18 04 08
	Liancalus	*virens* (Scopoli)	40 18 05 01
	Aphrosylus	*mitis* (Verrall)	40 18 06 01
	Porphyrops	*consobrina* (Zetterstedt)	40 18 07 01
		elegantula (Meigen)	40 18 07 02
		riparia Meigen	40 18 07 03
	Xiphandrium	*brevicorne* (Curtis)	40 18 08 01
	Syntormon	*filiger* Verrall	40 18 09 01
		mikii (Strobl)	40 18 09 02
		monile (Walker)	40 18 09 03
		pallipes (Fabricius)	40 18 09 04
		pumilum (Meigen)	40 18 09 05
		rufipes (Meigen)	40 18 09 06
		sulcipes (Meigen)	40 18 09 07
		tarsatum (Fallen)	40 18 09 08
		zelleri (Loew)	40 18 09 09
	Eutarsus	*aulicus* (Meigen)	40 18 10 01
	Machaerium	*maritimae* Haliday	40 18 11 01
	Systenus	*bipartitus* (Loew)	40 18 12 01
		leucurus (Loew)	40 18 12 02
		pallipes von Roser	40 18 12 03
		scholtzii (Loew)	40 18 12 04
		tener (Loew)	40 18 12 05
	Argyra	*argyria* (Meigen)	40 18 13 01
		confinis (Zetterstedt)	40 18 13 02
		leucocephala (Meigen)	40 18 13 03
	Leucostola	*vestita* (Wiedemann)	40 18 14 01
	Campsicnemus	*armatus* (Zetterstedt)	40 18 15 01
		curvipes (Fallen)	40 18 15 02
		loripes (Haliday)	40 18 15 03
		magius (Loew)	40 18 15 04
		pectinulatus (Loew)	40 18 15 05
		picticornis (Zetterstedt)	40 18 15 06
		pusillus (Meigen)	40 18 15 07
		scambus (Fallen)	40 18 15 08
	Teuchophorus	*monacanthus* (Loew)	40 18 16 01
		signatus (Staeger)	40 18 16 02
		simplex (Mik)	40 18 16 03
		spinigerellus Zetterstedt	40 18 16 04
	Anepsiomyia	*flaviventris* (Meigen)	40 18 17 01
Rhagionidae	*Atherix*	*ibis* (Fabricius)	40 19 01 01
Tabanidae	*Chrysops*	*caecutiens* (L.)	40 20 01 01
		relictus Meigen	40 20 01 02

		sepulcralis (Fabricius)	40 20 01 03
	Haematopota	*crassicornis* Wahlberg	40 20 02 01
		pluvialis (L.)	40 20 02 02
	Atylotus	*plebeius* (Fallen)	40 20 03 01
	Hybomitra	*bimaculata* (Macquart)	40 20 04 01
		schineri Lyneborg	40 20 04 02
	Tabanus	*autumnalis* L.	40 20 05 01
		sudeticus Zeller	40 20 05 02
		bromius L.	40 20 05 03
		maculicornis Zetterstedt	40 20 05 04
Syrphidae	*Eristalis*	*abusiva* Collin	40 21 01 01
		arbustorum L.	40 21 01 02
		horticola Degeer	40 21 01 03
		intricaria L.	40 21 01 04
		nemorum L.	40 21 01 05
		pertinax Scopoli	40 21 01 06
		rupium Fabricius	40 21 01 07
	Eristalomyia	*cryptarum* Fabricius	40 21 02 01
		tenax L.	40 21 02 02
	Lathrophthalmus	*aeneus* Scopoli	40 21 03 01
	Eristalinus	*sepulcralis* (L.)	40 21 04 01
	Myiatropa	*florea* L.	40 21 05 01
	Liops	*vittata* Meigen	40 21 06 01
	Tubifera	*groenlandica* Fabricius	40 21 07 01
		hybrida (Loew)	40 21 07 02
		pendula L.	40 21 07 03
		trivittata Fabricius	40 21 07 04
	Parhelophilus	*consimilis* Malm	40 21 08 01
		frutetorum Fabricius	40 21 08 02
		versicolor (Fabricius)	40 21 08 03
	Eurinomyia	*lineata* Fabricius	40 21 09 01
		lunulata Degeer	40 21 09 02
		transfuga L.	40 21 09 03
	Mallota	*cimbiciformis* Fallen	40 21 10 01
	Cinxia	*borealis* Fallen	40 21 11 01
		lappona L.	40 21 11 02
Ephydridae	*Ephydra*	*breviventris* Loew	40 22 01 01
		micans Haliday	40 22 01 02
		riparia Fallen	40 22 01 03
	Coenia	*fumosa* Stenhammar	40 22 02 01
		palustris Fallen	40 22 02 02
	Scatella	*paludum* Meigen	40 22 03 01
		stagnalis Fallen	40 22 03 02
	Halmopota	*salinarum* Bouche	40 22 04 01
	Ochthera	*mantis* Degeer	40 22 05 01
		mantispa Loew	40 22 05 02
	Dichaeta	*caudata* (Fallen)	40 22 06 01
	Notiphila	*annulipes* Stenhammar	40 22 07 01
		aquatica Becker	40 22 07 02
		brunnipes Robineau-Desvoidy	40 22 07 03
		cinerea Fallen	40 22 07 04
		dorsata Stenhammar	40 22 07 05
		maculata Stenhammar	40 22 07 06
		nigricornis Stenhammar	40 22 07 07
		riparia Meigen	40 22 07 08
		stagnicola Robineau-Desvoidy	40 22 07 09
		uliginosa Haliday	40 22 07 10

		venusta Loew	40 22 07 11
	Hydrellia	*albiceps* (Meigen)	40 22 08 01
		albilabris (Meigen)	40 22 08 02
		cardamines (Haliday)	40 22 08 03
		chrysostoma (Meigen)	40 22 08 04
		flavicornis (Fallen)	40 22 08 05
		griseola (Fallen)	40 22 08 06
		incana (Stenhammar)	40 22 08 07
		mutata (Zetterstedt)	40 22 08 08
		nigripes (Zetterstedt)	40 22 08 09
		ranunculi (Haliday)	40 22 08 10
Sciomyzidae	*Sepedon*	*sphegea* (Fabricius)	40 23 01 01
		spinipes Scopoli	40 23 01 02
	Dictya	*umbrarum* (L.)	40 23 02 01
	Elgiva	*cucularia* (L.)	40 23 03 01
		rufa (Panzer)	40 23 03 02
	Hydromya	*dorsalis* Fabricius	40 23 04 01
	Knutsonia	*albiseta* (Scopoli)	40 23 05 01
		lineata (Fallen)	40 23 05 02
	Limnia	*paludicola* Elberg	40 23 06 01
		unguicornis (Scopoli)	40 23 06 02
	Phorbina	*coryleti* (Scopoli)	40 23 07 01
	Psacadina	*punctata* (Fabricius)	40 23 08 01
		vittigera (Schiner)	40 23 08 02
	Tetanocera	*arrogans* (Meigen)	40 23 09 01
		ferruginea (Fallen)	40 23 09 02
		freyi Stack	40 23 09 03
		hyalipennis von Roser	40 23 09 04
		robusta Loew	40 23 09 05
		silvatica Meigen	40 23 09 06
		unicolor Loew	40 23 09 07
	Anticheta	*analis* (Meigen)	40 23 10 01
		brevipennis Zetterstedt	40 23 10 02
	Renocera	*pallida* (Fallen)	40 23 11 01
		striata (Meigen)	40 23 11 02
		strobli Hendel	40 23 11 03
	Colobaea	*bifasciella* (Fallen)	40 23 12 01
		distincta (Meigen)	40 23 12 02
		pectoralis (Zetterstedt)	40 23 12 03
		punctata (Lundbeck)	40 23 12 04
	Pherbellia	*brunnipes* (Meigen)	40 23 13 01
		cinerella (Fallen)	40 23 13 02
		griseola (Fallen)	40 23 13 03
		grisescens (Meigen)	40 23 13 04
		lata (Schiner)	40 23 13 05
		nana (Fallen)	40 23 13 06
		obtusa (Fallen)	40 23 13 07
		schoenherri (Fallen)	40 23 13 08
		ventralis (Fallen)	40 23 13 09
	Pteromicra	*glabricula* (Fallen)	40 23 14 01
		leucopeza (Meigen)	40 23 14 02
		nigrimana (Meigen)	40 23 14 03
		pectorosa (Hendel)	40 23 14 04
	Sciomyza	*simplex* Fallen	40 23 15 01
Scatophagidae	*Acanthocnema*	*nigrimana* (Zetterstedt)	40 24 01 01
		glaucescens (Loew)	40 24 01 02
	Spaziphora	*hydromyzina* (Fallen)	40 24 02 01
	Hydromyza	*livens* (Fabricius)	40 24 03 01

Muscidae	*Phaonia*	*exoleta* (Meigen)	40 25 01 01
	Limnophora	*riparia* (Fallen)	40 25 02 01
	Lispe	*caesia* Meigen	40 25 03 01
		consanguinea Loew	40 25 03 02
		pygmaea Fallen	40 25 03 03
		tentaculata (Degeer)	40 25 03 04
		uliginosa (Fallen)	40 25 03 05
41 VERTEBRATA	AGNATHA		
Petromyzonidae	*Petromyzon*	*marinus* L.	41 01 01 01
	Lampetra	*fluviatilis* (L.)	41 01 02 01
		planeri (Bloch)	41 01 02 02
	PISCES		
Acipenseridae	*Acipenser*	*sturio* L.	41 02 01 01
Clupeidae	*Alosa*	*alosa* (L.)	41 03 01 01
		fallax (Lacepede)	41 03 01 02
Salmonidae	*Salmo*	*salar* L.	41 04 01 01
		trutta L.	41 04 01 02
		gairdneri Richardson	41 04 01 03
	Oncorhynchus	*gorbuscha* (Walbaum)	41 04 02 01
	Salvelinus	*alpinus* (L.)	41 04 03 01
		fontinalis (Mitchill)	41 04 03 02
Coregonidae	*Coregonus*	*lavaretus* (L.)	41 05 01 01
		albula (L.)	41 05 01 02
		oxyrinchus (L.)	41 05 01 03
Thymallidae	*Thymallus*	*thymallus* (L.)	41 06 01 01
Osmeridae	*Osmerus*	*eperlanus* (L.)	41 07 01 01
Esocidae	*Esox*	*lucius* L.	41 08 01 01
Cyprinidae	*Cyprinus*	*carpio* L.	41 09 01 01
	Carassius	*carassius* (L.)	41 09 02 01
		auratus (L.)	41 09 02 02
	Barbus	*barbus* (L.)	41 09 03 01
	Gobio	*gobio* (L.)	41 09 04 01
	Tinca	*tinca* (L.)	41 09 05 01
	Blicca	*bjoerkna* (L.)	41 09 06 01
	Abramis	*brama* (L.)	41 09 07 01
	Alburnus	*alburnus* (L.)	41 09 08 01
	Phoxinus	*phoxinus* (L.)	41 09 09 01
	Rhodeus	*sericeus* (Bloch)	41 09 10 01
	Scardinius	*erythrophthalmus* (L.)	41 09 11 01
	Rutilus	*rutilus* (L.)	40 09 12 01
	Leuciscus	*cephalus* (L.)	40 09 13 01
		idus (L.)	40 09 13 02
		leuciscus (L.)	40 09 13 03
Cobitidae	*Cobitis*	*taenia* L.	41 10 01 01
	Noemacheilus	*barbatulus* (L.)	41 10 02 01
Siluridae	*Silurus*	*glanis* L.	41 11 01 01
Anguillidae	*Anguilla*	*anguilla* (L.)	41 12 01 01
Gasterosteidae	*Gasterosteus*	*aculeatus* L.	41 13 01 01
	Pungitius	*pungitius* (L.)	41 13 02 01
Gadidae	*Lota*	*lota* (L.)	41 14 01 01
Serranidae	*Dicentrarchus*	*labrax* (L.)	41 15 01 01
Centrarchidae	*Micropterus*	*salmoides* (Lacepede)	41 16 01 01
	Lepomis	*gibbosus* (L.)	41 16 02 01

	Ambloplites	*rupestris* (Rafinesque-Schmaltz)	41 16 03 01
Percidae	*Perca*	*fluviatilis* L.	41 17 01 01
	Gymnocephalus	*cernua* (L.)	41 17 02 01
	Stizostedion	*lucioperca* (L.)	41 17 03 01
Gobiidae	*Pomatoschistus*	*microps* (Kroyer)	41 18 01 01
Mugilidae	*Crenimugil*	*labrosus* (Risso)	41 19 01 01
	Chelon	*ramada* (Risso)	41 19 02 01
		auratus (Risso)	41 19 02 02
Cottidae	*Cottus*	*gobio* L.	41 20 01 01
Pleuronectidae	*Platichthys*	*flesus* (L.)	41 21 01 01

42 AMPHIBIA

Salamandridae	*Triturus*	*cristatus* (Laurenti)	42 01 01 01
		vulgaris (L.)	42 01 01 02
		helveticus Razoumoski	42 01 01 03
Bufonidae	*Bufo*	*bufo* (L.)	42 02 01 01
		calamita Laurenti	42 02 01 02
Ranidae	*Rana*	*temporaria* L.	42 03 01 01
		esculenta L.	42 03 01 02
		ridibunda Pallas	42 03 01 03

43 REPTILIA

Colubridae	*Natrix*	*natrix* (L.)	43 01 01 01
	Coronella	*austriaca* Laurenti	43 01 01 02

44 AVES

Gaviidae	*Gavia*	*arctica* (L.)	44 01 01 01
		immer (Brunnich)	44 01 01 02
		adamsii (Gray)	44 01 01 03
		stellata (Pontoppidan)	44 01 01 04
Podicipidae	*Podiceps*	*cristatus* (L.)	44 02 01 01
		griseigena (Boddaert)	44 02 01 02
		auritus (L.)	44 02 01 03
		nigricollis Brehm	44 02 01 04
	Tachybaptus	*ruficollis* (Pallas)	44 02 02 01
	Podilymbus	*podiceps* (L.)	44 02 03 01
Procellariidae	*Fulmarus*	*glacialis* (L.)	44 03 01 01
Phalacrocoracidae	*Phalacrocorax*	*carbo* (L.)	44 04 01 01
		aristotelis (L.)	44 04 01 02
Ardeidae	*Ardea*	*cinerea* L.	44 05 01 01
		purpurea L.	44 05 01 02
	Egretta	*garzetta* (L.)	44 05 02 01
		alba (L.)	44 05 02 02
	Ardeola	*ralloides* (Scopoli)	44 05 03 01
	Bubulcus	*ibis* (L.)	44 05 04 01
	Nycticorax	*nycticorax* (L.)	44 05 05 01
	Ixobrychus	*minutus* (L.)	44 05 06 01
	Botaurus	*stellaris* (L.)	44 05 07 01
		lentiginosus (Montagu)	44 05 07 02
Ciconiidae	*Ciconia*	*ciconia* (L.)	44 06 01 01
		nigra (L.)	44 06 01 02
Threskiornithidae	*Platalea*	*leucorodia* L.	44 07 01 01
	Plegadis	*falcinellus* (L.)	44 07 02 01
Anatidae	*Anas*	*platyrhynchos* L.	44 08 01 01
		rubripes Brewster	44 08 01 02
		crecca L.	44 08 01 03

		querquedula L.	44 08 01 04
		discors L.	44 08 01 05
		strepera L.	44 08 01 06
		penelope L.	44 08 01 07
		americana Gmelin	44 08 01 08
		acuta L.	44 08 01 09
		clypeata (L.)	44 08 01 10
	Aix	*galericulata* (L.)	44 08 02 01
	Netta	*rufina* (Pallas)	44 08 03 01
	Aythya	*marila* (L.)	44 08 04 01
		fuligula (L.)	44 08 04 02
		collaris (Donovan)	44 08 04 03
		ferina (L.)	44 08 04 04
		nyroca (Guldenstadt)	44 08 04 05
	Bucephala	*albeola* (L.)	44 08 05 01
		clangula (L.)	44 08 05 02
	Clangula	*hyemalis* (L.)	44 08 06 01
	Melanitta	*fusca* (L.)	44 08 07 01
		perspicillata (L.)	44 08 07 02
		nigra (L.)	44 08 07 03
	Histrionicus	*histrionicus* (L.)	44 08 08 01
	Polysticta	*stelleri* (Pallas)	44 08 09 01
	Somateria	*mollissima* (L.)	44 08 10 01
		spectabilis (L.)	44 08 10 02
	Oxyura	*jamaicensis* (Gmelin)	44 08 11 01
	Mergus	*serrator* L.	44 08 12 01
		merganser L.	44 08 12 02
		albellus L.	44 08 12 03
		cucullatus L.	44 08 12 04
	Tadorna	*tadorna* (L.)	44 08 13 01
		ferruginea (Pallas)	44 08 13 02
	Alopochen	*aegyptiacus* (L.)	44 08 14 01
	Anser	*anser* (L.)	44 08 15 01
		albifrons (Scopoli)	44 08 15 02
		erythropus (L.)	44 08 15 03
		fabalis (Latham)	44 08 15 04
		brachyrhynchus Baillon	44 08 15 05
		caerulescens (L.)	44 08 15 06
	Branta	*bernicla* (L.)	44 08 16 01
		leucopsis (Bechstein)	44 08 16 02
		canadensis (L.)	44 08 16 03
		ruficollis (Pallas)	44 08 16 04
	Cygnus	*olor* (Gmelin)	44 08 17 01
		cygnus (L.)	44 08 17 02
		bewickii Yarrell	44 08 17 03
Pandionidae	*Pandion*	*haliaetus* (L.)	44 09 01 01
Gruidae	*Grus*	*grus* (L.)	44 10 01 01
Rallidae	*Rallus*	*aquaticus* L.	44 11 01 01
	Porzana	*porzana* (L.)	44 11 02 01
		carolina (L.)	44 11 02 02
		pusilla (Pallas)	44 11 02 03
		parva (Scopoli)	44 11 02 04
	Gallinula	*chloropus* (L.)	44 11 03 01
	Porphyrula	*martinica* (L.)	44 11 04 01
	Fulica	*atra* L.	44 11 05 01
Haematopodidae	*Haematopus*	*ostralegus* L.	44 12 01 01
Charadriidae	Vanellus	*gregarius (Pallas)*	44 13 01 01

		vanellus (L.)	44 13 01 02
	Charadrius	hiaticula L.	44 13 02 01
		dubius Scopoli	44 13 02 02
		alexandrinus L.	44 13 02 03
		vociferus L.	44 13 02 04
		asiaticus Pallas	44 13 02 05
	Pluvialis	squatarola (L.)	44 13 03 01
		apricaria (L.)	44 13 03 02
		dominica (Muller)	44 13 03 03
	Eudromias	morinellus (L.)	44 13 04 01
Scolopacidae	Arenaria	interpres (L.)	44 14 01 01
	Limnodromus	griseus (Gmelin)	44 14 02 01
		scolopaceus (Say)	44 14 02 02
	Micropalama	himantopus (Bonaparte)	44 14 03 01
	Gallinago	gallinago (L.)	44 14 04 01
		media (Latham)	44 14 04 02
	Lymnocryptes	minimus (Brunnich)	44 14 05 01
	Scolopax	rusticola L.	44 14 06 01
	Bartramia	longicauda (Bechstein)	44 14 07 01
	Numenius	arquata (L.)	44 14 08 01
		phaeopus (L.)	44 14 08 02
		borealis (Forster)	44 14 08 03
	Limosa	limosa (L.)	44 14 09 01
		lapponica (L.)	44 14 09 02
	Tringa	ochropus L.	44 14 10 01
		glareola L.	44 14 10 02
		solitaria Wilson	44 14 10 03
		hypoleucos (L.)	44 14 10 04
		macularia (L.)	44 14 10 05
		totanus (L.)	44 14 10 06
		erythropus (Pallas)	44 14 10 07
		melanoleuca (Gmelin)	44 14 10 08
		flavipes (Gmelin)	44 14 10 09
		nebularia (Gunnerus)	44 14 10 10
		stagnatilis (Bechstein)	44 14 10 11
	Xenus	cinereus (Guldenstadt)	44 14 11 01
	Calidris	canutus (L.)	44 14 12 01
		maritima (Brunnich)	44 14 12 02
		minuta (Leisler)	44 14 12 03
		minutilla (Vieillot)	44 14 12 04
		temminckii (Leisler)	44 14 12 05
		bairdii (Coues)	44 14 12 06
		fuscicollis (Vieillot)	44 14 12 07
		melanotos (Vieillot)	44 14 12 08
		acuminata (Horsfield)	44 14 12 09
		alpina (L.)	44 14 12 10
		ferruginea (Pontoppidan)	44 14 12 11
		pusilla (L.)	44 14 12 12
		mauri (Cabanis)	44 14 12 13
		alba (Pallas)	44 14 12 14
	Trynigites	subruficollis (Vieillot)	44 14 13 01
	Limicola	falcinellus (Pontoppidan)	44 14 14 01
	Philomachus	pugnax (L.)	44 14 15 01
Recurvirostridae	Recurvirostra	avosetta L.	44 15 01 01
	Himantopus	himantopus (L.)	44 15 02 01
Phalaropodidae	Phalaropus	fulicarius (L.)	44 16 01 01
		lobatus (L.)	44 16 01 02
		tricolor (Vieillot)	44 16 01 03

Burhinidae	*Burhinus*	*oedicneumus* (L.)	44 17 01 01
Stercorariidae	*Stercorarius*	*skua* (Brunnich)	44 18 01 01
		pomarinus (Temminck)	44 18 01 02
		parasiticus (L.)	44 18 01 03
		longicaudus Vieillot	44 18 01 04
Laridae	*Pagophila*	*eburnea* (Phipps)	44 19 01 01
	Larus	*marinus* L.	44 19 02 01
		fuscus L.	44 19 02 02
		argentatus Pontoppidan	44 19 02 03
		canus L.	44 19 02 04
		hyperboreus Gunnerus	44 19 02 05
		glaucoides Meyer	44 19 02 06
		genei Breme	44 19 02 07
		ichthyaetus Pallas	44 19 02 08
		melanocephalus Temminck	44 19 02 09
		atricilla L.	44 19 02 10
		philadelphia (Ord)	44 19 02 11
		minutus Pallas	44 19 02 12
		ridibundus L.	44 19 02 13
		sabini Sabine	44 19 02 14
	Rhodostethia	*rosea* (MacGillivray)	44 19 03 01
	Rissa	*tridactyla* (L.)	44 19 04 01
	Chlidonias	*niger* (L.)	44 19 05 01
		leucopterus (Temminck)	44 19 05 02
		hybrida (Pallas)	44 19 05 03
	Gelochelidon	*nilotica* (Gmelin)	44 19 06 01
	Hydroprogne	*caspia* (Pallas)	44 19 07 01
	Sterna	*hirundo* L.	44 19 08 01
		paradisaea Pontoppidan	44 19 08 02
		dougallii Montagu	44 19 08 03
		fuscata L.	44 19 08 04
		anaethetus Scopoli	44 19 08 05
		albifrons Pallas	44 19 08 06
		maxima Boddaert	44 19 08 07
		sandvicensis Latham	44 19 08 08
Alcedinidae	*Alcedo*	*atthis* (L.)	44 20 01 01
Cinclidae	*Cinclus*	*cinclus* (L.)	44 21 01 01
Timaliidae	*Panurus*	*biarmicus* (L.)	44 22 01 01
Sylviidae	*Cettia*	*cetti* (Temminck)	44 23 01 01
	Locustella	*naevia* (Boddaert)	44 23 02 01
		fluviatilis (Wolf)	44 23 02 02
		luscinioides (Savi)	44 23 02 03
	Acrocephalus	*melanopogon* (Temminck)	44 23 03 01
		arundinaceus (L.)	44 23 03 02
		scirpaceus (Hermann)	44 23 03 03
		palustris (Bechstein)	44 23 03 04
		dumetorum (Blyth)	44 23 03 05
		agricola (Jerdon)	44 23 03 06
		schoenobaenus (L.)	44 23 03 07
		paludicola (Vieillot)	44 23 03 08
	Hippolais	*polyglotta* (Vieillot)	44 23 04 01
		icterina (Vieillot)	44 23 04 02
		pallids (Hemrpich&Ehrenberg)	44 23 04 03
		caligata (Lichtenstein)	44 23 04 04
Motacillidae	*Anthus*	*spinoletta* (L.)	44 24 01 01
	Motacilla	*alba* L.	44 24 02 01
		cinerea Tunstall	44 24 02 02
		citreola Pallas	44 24 02 03
		flava L.	44 24 02 04

Emberizidae	*Emberiza*	*schoeniclus* (L.)	44 25 01 01
45	MAMMALIA		
Soricidae	*Neomys*	*fodiens* (Pennant)	45 01 01 01
Mustelidae	*Mustela*	*vison* Schreber	45 02 01 01
	Lutra	*lutra* (L.)	45 02 01 01
Phocidae	*Halichoerus*	*grypus* (Fabricius)	45 03 01 01
	Phoca	*vitulina* L.	45 03 02 01
Muridae	*Arvicola*	*terrestris* (L.)	45 04 01 01
	Ondatra	*zibethicus* (L.)	45 04 02 01
Capromyidae	*Myocastor*	*coypus* (Molina)	45 05 01 01

References

1. ANDRASSY, I. 1967. Nematoda (Errantia). *Limnofauna Europaea,* 1, 73-88.

2. BALFOUR-BROWNE, F. 1964. *British water beetles.* Ray Society, London.

3. BALOGH, J. 1963. Identification keys of holarctic oribatid mites (Acari) families and genera. *Acta zool., Budapest,* 9, 1-60.

4. BAYLIS, H.A. 1943. Notes on the distribution of hairworms (Nematomorpha: Gordiidae) in the British Isles. *Proc. Zool. Soc. Lond., B, 113, 193-197.*

5. BECKER, T. 1926. *Die Fliegen der Palaearktischen Region. 56a. Ephydridae.* Schweizerbart. Stuttgart:

6. BERZINS, B.V.A. 1967. Rotatoria. *Limnofauna Europaea,* 1, 35-68.

7. BRINKHURST, R.O. 1961. A check list of British Oligochaeta. *Proc. zool. Soc. Lond.,* 138, 317-330.

8. BRINKHURST, R.O. 1963. A guide for the identification of British aquatic Oligochaeta. *Sci. Publ. Freshw. Biol., Ass.,* 22, 1-52.

9. BRITISH ORNITHOLOGISTS' UNION 1971. *The status of birds in Britain & Ireland.* Blackwell, Oxford.

10. BRITISH TRUST FOR ORNITHOLOGY, 1971. A species list of British and Irish birds. *Brit. Trust Ornith. Guide,* 13, 1-18.

11. BRONSTEIN, Z.S. 1947. Ostracodes des eaux douces. Faune de l'URSS. Crustaces. *Inst. Zool. Acad. Sci. URSS.,* 31, 1-339.

12. CAMPBELL, J.A. & PELHAM-CLINTON, E.C. 1960. A taxonomic review of the British species of *Culicoides* Latreille (Diptera Ceratopogonidae). *Proc. R. Soc. Edin.,* 67, B, 14-302.

13. COE, R.L. 1950. Diptera: Family Tipulidae. *R. Ent. Soc. Lond., Handb. Ident. Brit. Ins.,* 9, 121-206.

14. COE, R.L. 1950. Diptera: Family Chironomidae. *R. Ent. Soc. Lond., Handb. Ident. Brit. Ins.,* 9, 121-206.

15. COE. R.L. 1953. Diptera: Syrphidae. *R. Ent. Soc. Lond., Handb. Ident. Brit. Ins.,* 10, 1-98.

16. COLLIN, J.E. 1958. A short synopsis of the British Scatophagidae (Diptera) *Trans. Soc. Brit. Ent.,* 13, 37-56.

17. COLLIN, J.E. 1961. *British Flies: VI: Empididae.* Cambridge

18. CORBET, G.B. 1969. The identification of British mammals. *Publ. Brit. Mus. Nat. Hist. Lond.* 623, 1-46.

19. CORBET, P.S., LONGFIELD, C., & MOORE, N.W. 1960. *Dragonflies.* Collins: London.

20. DAHL, R. 1967. Insecta: Diptera: Ephydridae. *Limnofauna Europaea,* 1, 415-416.

21. D'ASSIS-FONSECA. 1968. Diptera: Cylorrhapha (Muscidae). *R. Ent. Soc. Lond., Handb. Ident. Brit. Ins.,* 10, 1-119.

22. DAVIES, L. 1968. A key to the British species of Simuliidae (Diptera) in the larval, pupal and adult stages. *Sci. Publ. Freshw. Biol. Ass.,* 24, 1-126.

23. DAWES, B. 1946. *The Trematoda with special reference to British and other European forms.* University Press, Cambridge.

24. DISNEY, R.H.L. 1975. A key to the larvae, pupae and adults of the British Dixidae (Diptera), the meniscus midges. *Sci. Publ. Freshw. Biol. Ass.,* 31, 1-78.

25. EALES, N.B. 1950. *The littoral fauna of Great Britain.* University Press, Cambridge.

26. EDWARDS, F.W. 1929. Revision of Thaumaleidae. *Zool. Anz.,* 82, 121-142.

27. ELLIS, A.E. 1951. Census of the distribution of British non-marine Mollusca. *J. Conch., Lond.,* 23, 171-246.

28. ELLIS, A.E. 1962. British freshwater bivalve molluscs, with keys and notes for the identification of the species. *Linn. Soc. Lond., Synop. Brit. Fauna* 13, 1-92.

29. FITTER, R.S.R. & RICHARDSON, R.A. 1952. *Pocket guide to British Birds. The complete identification book.* Collins, London.

30. FRASER, F.C. 1956. Odonata. *Roy. Ent. Soc. Lond. Handb. Ident. Brit. Ins.,* 1, 1-49.

31. FREEMAN, P. 1950. Diptera: Ptychopteridae. *R.Ent. Soc. Lond., Handb. Ident. Brit. Ins.,* 9, 73-76.

32. FREEMAN, P. 1950. Diptera: Family Psychodidae. *R. Ent. Soc. Lond. Handb. Ident. Brit. Ins.,* 9, 77-96.

33. FREEMAN, P. 1950. Diptera: Family Culicidae: Subfamilies Dixinae and Chaoborinae. *R. Ent. Soc. Lond., Handb. Ident. Brit. Ins.,* 9, 97-101.

34. GERARD, B.M. 1964. A synopsis of the British Lumbricidae. *Linn. Soc. Lond., Synops, Brit. Fauna,* 6, 1-58.

35. GIBSON, R. & MOORE, J. 1977. *Freshwater nemerteans.* In preparation.

36. GIBSON, R. & YOUNG, J.O. 1971. *Prostoma jenningsi* sp. nov., a new British freshwater hoplonemertean. *Freshw. Biol.,* 1, 121-127.

37. GISIN, H. 1960 *Collembolen fauna Europas.* Museum d'Histoire Naturelle, Geneva.

38. GISIN, H. 1967. Insecta: Collembola. *Limnofauna Europaea,* 1, 210-211.

39. GLEDHILL, T. 197. A check-list of the freshwater mites (Hydrachnellae and Limnohalacaridae, Acari) recorded from Great Britain and Ireland. Occas. Publ. Freshw. Biol. Ass., 1, 1-59.

40. GLEDHILL, T., SUTCLIFFE, D.W. & WILLIAMS, W.D. 1976. A revised key to the British species of Crustacea: Malacostraea occurring in fresh water. *Sci. Publ. Freshw. Biol. Ass.,* 31, 1-72.

41. GOODEY, T. 1963. *Soil and freshwater nematodes.* Methuen, London.

42. GRAYSON, R.F. 1971. The freshwater Hydras of Europe. 1. A review of the European species. *Arch. Hydrobiol.,* 68, 436-449.

43. GURNEY, R. 1933. *British freshwater Copepoda.* Ray Society: London

44. GURNEY, R. 1948. British species of the fish louse of the genus *Argulus. Proc. zool. Soc. Lond.,* 118, 553-558.

45. HANNEMANN, H. 1967. Insecta: Lepidoptera. *Limnofauna Europaea,* 1, 310-311.

46. HARDING, J.P. & SMITH, W.A. 1960. A key to the British freshwater cyclopid and calanoid copepods. *Sci. Pub. Freshw. Biol. Ass.* 18, 1-54.

47. HEDQVIST, K. 1967. Insecta: Hymenoptera. *Limnofauna Europaea,* 1, 242-244.

48. HENNIG, W. 1967. Insecta: Diptera: Muscidae. *Limnofauna Europaea,* 1, 423-424.

49. HICKIN, N.E. 1967. *Caddis larvae: larvae of the British Trichoptera.* Hutchison, London.

50. HOLLAND, D.G. 1972. A key to the larvae pupae and adults of the British species of Elminthidae. *Sci. Publ. Freshw. Biol. Ass.,* 26, 1-58.

51. HORKAN, J.P.K. 1975. A list of the Rotatoria known to occur in Ireland, with notes on their habitats and distribution. Unpublished MS.

52. HYMAN, L.H. 1951. *The invertebrates: Acanthocephala, Aschelminthes and Entoprocta.* McGraw-Hill: New York.

53. HYNES, H.B.N. 1967. A key to the British adults and nymphs of stoneflies (Plecoptera). *(Sci. Publ. Freshw. Biol. Ass.* 17, 1-86.

54. ILLIES, J. 1967. *Limnofauna Europaea.* Fischer: Stuttgart.

55. JOY, N.H. 1975. *A practical handbook of British beetles.* Witherby, London.

56. KAURI, H. 1967. Insecta: Diptera: Tabanidae et Leptidae. *Limnofauna Europaea,* 1, 410-411.

57. KERRICH, G.J. 1960. The state of our knowledge of the systematics of the Hymenoptera Parasitica, with particular reference to the British fauna. *Trans. Soc. Brit. Ent.,* 14, 1-18.

58. KIEFER, F. 1967. Branchiura. *Limnofauna Europaea,* 1, 186.

59. KIMMINS, D.E. 1962. Keys to the British species of aquatic Megaloptera and Neuropera with ecological notes. *Sci. Publ. Freshw. Biol. Ass.,* 8, 1-23.

60. KIMMINS, D.E. 1966. A revised check-list of the British Trichoptera. *Ent. Gaz.* 17, 111-120.

61. KIMMINS, D.E. 1972. A revised key to the adults of the British species of Ephemeroptera with notes on their ecology. *Sci. Publ. Freshw. Biol. Ass.* 15, 1-75.

62. KLOET, G.S. & HINCKS,W.D. 1964. A check list of British insects. Part 1: small orders and Hemiptera. *R. Ent. Soc. Lond., Handb. Ident. Brit. Ins.* 11,1-119.

63. KLOET, G.S. & HINCKS, W.D. 1972. A check list of British insects. Part 2: Lepidoptera *R. Ent. Soc. Lond., Handb. Ident. Brit. Ins.,* 11, 1-153.

64. KLOET, G.S. & HINCKS, W.D. 1975. A check list of British insects. Part 3: Diptera. *R. Ent. Soc. Lond., Handb. Ident. Brit. Ins.,* 11, in press.

65. LACOUR, A.W. 1968. A monograph of the freshwater Bryozoa-Phylactolaemata. *Zool. verhand.,* 93, 1-159.

66. LANG, K. 1948. *Monographie der Harpacticiden.* Ohlssons, Lund.

67. LE GROS, A.E. 1958. How to begin the study of tardigrades. *Countryside,* 18, 1-11.

68. LOCKET, G.M., MILLIDGE, A.F. & MERRETT, P. 1974. *British spiders.* Ray Society London.

69. LOFFLER, M. 1967. Ostracoda. *Limnofauna Europaea,* 1, 162-172.

70. LOFFLER, H. 1967. Anostraca, Notostraca, Conchostraca. *Limnofauna Europaea,* 1, 151-155.

71. MACAN, T.T. 1961. A key to the nymphs of the British species of Ephemeroptera. *Sci. Publ. Freshw. Biol. Ass.* 20. 1-63.

72. MACAN, T.T. 1965. A revised key to the British water bugs. (Hemiptera-Heteroptera). *Sci. Publ. Freshw. Biol. Ass.* 16, 1-78.

73. MACAN, T.T. 1969. A key to the British fresh-and brackish-water gastropods with notes on their ecology. *Sci. Publ. Freshw. Biol. Ass.,* 13, 1-46.

74. MACAN, T.T. & WORTHINGTON, C.J. 1973. A key to the adults of the British Trichoptera. *Sci. Publ. Freshw. Biol. Ass.,* 28, 1-151.

75. McMILLAN, N.F. 1968. *British shells.* Warne: London.

76. MAITLAND, P.S. 1972. A key to the freshwater fishes of the British Isles with notes on their distribution and ecology. *Sci. Publ. Freshw. Biol. Ass.,* 27, 1-139.

77. MANN, K.H. 1964. A key to the British freshwater leeches with notes on their ecology. *Sci. Publ. Freshw. Biol. Ass.,* 14, 1-50.

78. MATTINGLY, P.F. 1950. Diptera: Family Culicidae: Subfamily Culicinae. *R. Ent. Soc. Lond., Handb. Ident. Brit. Ins.,* 9, 102-120.

79. MEYRICK, E. 1928. *A revised handbook of the British Lepidoptera.* London, Watkins & Doncaster.

80. MURRAY, J. 1911. Scottish Tardigrada. A Review of our present knowledge. *Ann. Scot. Nat. Hist.* 88-95.

81. NIELSEN, C.O. & CHRISTENSEN, B. 1959. The Enchytraeidae: critical revision and taxonomy of European species. *Natura Jutlandica,* 8, 1-159; (1961) 10, 1-23; (1963) 12, 1-19.

82. OLDROYD, H. 1954. Diptera: introduction and keys to families. *R. Ent. Soc. Lond., Handb. Ident. Brit. Ins.,* 9, 1-49.

83. OLDROYD, H. 1969. Diptera: Brachycera, section (a) Tabanoidea and Asiloidea. *R. Ent. Soc. Lond., Handb. Ident. Brit. Ins.,* 9, 1-132.

84. PARENT, O. 1938. *Faune de France. 35. Dolichopodidae.* Lechevalier: Paris.

85. PARMENTER, L. 1966. The Sciomyzidae (Diptera) in Britain. *Ent. Record,* 78, 125-128.

86. PENNEY, J.T. & RACEK, A.A. 1968. Comprehensive revision of a worldwide collection of freshwater sponges (Porifera: Spongillidae). *Bull. U.S. natn. Mus.,* 272, 1-184.

87. PIFFL, E. 1967. Arachnida: Oribatei. *Limnofauna Europaea,* 1, 149-150.

88. PONTIN, R.M. 1976. A key to British Rotifera. In press.

89. RAMAZZOTTI, G. 1967. Tardigrada. *Limnofauna Europaea,* 1, 121-123.

90. REMANE, A. 1967. Gastrotricha. *Limnofauna Europaea,* 1, 69-72.

91. REYNOLDSON, T.B. 1967. A key to the British species of freshwater triclads. *Sci. Publ. Freshw. Biol. Ass.,* 23, 1-28.

92. RICHARDS. O.W. 1956. Hymenoptera. Introduction and keys to families. *R. Ent. Soc. Lond., Handb. Ident. Brit. Ins.,* 6, 1-94.

93. RITCHIE, J. 1915. Scottish hairworms . . . . with a key for the discrimination of the species recorded from Britain. *Scot. Nat.,* 111-115, 136-142, 255-262.

94. RUTTNER-KOLISKO, A. 1974. Plankton rotifers: biology and taxonomy. *Die Binnengewasser,* 26, 1-146.

95. SAETHER, O.A. 1970. Nearctic and Palaearctic *Chaoborus* (Diptera: Chaoboridae). *Bull. Fish. Res. Bd. Canada,* 174, 1-75.

96. SCHULZ, E. 1967. Hydrozoa. *Limnofauna Europaea,* 1, 3-4.

97. SCOURFIELD, D.J. & HARDING, J.P. 1966. A key to the British species of freshwater Cladocera with notes on their ecology. *Sci. Publ. Freshw. Biol. Ass.,* 5, 1-55.

98. SMITH, I.R. 1976. An index of British lakes and reservoirs. In preparation.

99. SMITH, M.A. 1964. *British reptiles and Amphibia.* Collins, London.

100. SOAR, C.D. & WILLIAMSON, W. 1929. *British Hydracarina.* Ray Society, London.

101. STEPHENS, J. 1920. The freshwater sponges of Ireland. *Proc. R. Irish Acad.,* B, 35, 205-254.

102. TYNEN, M.J. 1966. A new species of *Lumbricillus* with a revised check-list of the British Enchytraeidae (Oligochaeta). *J. mar. biol. Ass.,* 46, 89-95.

103. VAILLANT, F. 1967. Insecta: Diptera: Empididae. *Limnofauna Europaea,* 1, 401-404.

104. VAILLANT, F. 1967. Insecta: Diptera: Dolichopodidae. *Limnofauna Europaea,* 1, 405-409.

105. VENTURI, F. 1967. Insecta: Diptera: Syrphidae. *Limnofauna Europaea,* 1, 412-414.

106. VERBEKE, J.L. & KNUTSON, L.V. 1967. Insecta: Diptera: Sciomyzidae. *Limnofauna Europaea,* 1, 417-421.

107. VERRAL, G.H. 1909. *British flies V. Stratiomyidae.* Gurney & Jackson, London.

108. VOCKEROTH, J.R. 1967. Insecta: Diptera: Scatophagidae. *Limnofauna Europaea,* 1, 422.

109. VOIGT, M. 1957. *Rotatoria. Die Radertiere Mitteleuropas.* Borntrager: Berlin.

110. VOIGT, M. 1958. Gastrotricha Gastrotrichen. *Tierwelt Mitteleuropas.* 1, 4a, 1-74.

111. WIEBACH, F. 1967. Bryozoa. *Limnofauna Europaea,* 1, 425-426.

112. YAMAGUTI, S. 1959. *Systema helminthum, II, the cestodes of vertebrates.* Interscience, New York.

113. YOUNG, J.O. 1970. British and Irish freshwater Microturbellaria: historical records, new records and a key for their identification. *Arch. Hydrobiol.,* 67, 210-241.

114. YOUNG, J.O. 1972. Further studies on the occurrence of freshwater Microturbellaria in the British Isles II. New records and an emendation to the existing key for the group. *Freshw. Biol.,* 2, 355-359.